Quantity Surveying Practice

Quantity Surveying Practice: The Nuts and Bolts is a practical guide to quantity surveying in building construction. Due to the increasing expectations of quality and performance from project clients, quantity surveyors must improve their professional skills to solve a variety of intricate problems and disputes confronting the demanding construction market. This practical book focuses on the basic concepts underlying the technical aspects of quantity surveying and contains many worked examples together with useful figures and real-life cases to help readers digest and understand the essentials and become better professionals as a result.

This book is organised and structured into seven chapters. Chapter 1 is about the estimation of construction costs. Chapter 2 gives an overview of tendering and tender documentation. Chapter 3 examines the procedure of tender examination and the approach to contract award. Chapter 4 reviews the whole process of an interim valuation from the submission of a payment application by the contractor to the issuance of an interim valuation by the quantity surveyor, identifying the key issues within the process. Chapter 5 examines the topic of construction claims. Chapter 6 addresses the cost control and monitoring in connection with construction projects. Chapter 7 is about dispute management and three commonly used dispute resolution mechanisms, namely mediation, adjudication and arbitration are introduced.

This book is essential reading for students on quantity surveying and construction management programmes, as well as the APC candidates pursuing the professional quantity surveying pathway. It is also a useful reference for practicing quantity surveyors.

Sr Chung Wai Calvin Keung PhD, FHKIS, FRICS
Dr Keung practiced as a consultant quantity surveyor in Hong Kong before moving to academia in 2006. He is one of the core academic staff teaching quantity surveying and building information modelling (BIM) at the City University of Hong Kong. Dr Keung's current research interests include 5D BIM solutions and BIM-enabled virtual design and collaboration. Dr Keung was awarded the BIMers 2020 by the Construction Industry Council.

Sr Kam Lan Daisy Yeung PhD, MHKIS, FRICS
Dr Yeung has taught at the City University of Hong Kong since 2002 and is currently the programme leader of the Bachelor of Science in Surveying. With over 20 years' experience of quantity surveying practice in the construction industry, Dr Yeung has worked for cost consultant firms, property developer companies and organisations in the public sector. Dr Yeung has conducted research in construction cost forecasting, construction contracts and cost control performance.

Sr Sai On Cheung DSc, PhD, FHKIS, FRICS
Dr Cheung engaged in quantity surveying practice in both contracting and consultancy offices before joining the City University of Hong Kong in 1989. Dr Cheung is the subject leader for legal studies in the City University's Surveying programme and the head of the Construction Dispute Resolution Research Unit (CDRRU) of the Department of Architecture and Civil Engineering. The CDRRU has conducted a series of successful research projects covering organisational issues in construction with a focus on construction disputes.

Quantity Surveying Practice

The Nuts and Bolts

Chung Wai Calvin Keung,
Kam Lan Daisy Yeung and
Sai On Cheung

LONDON AND NEW YORK

First published 2022
by Routledge
2 Park Square, Milton Park, Abingdon, Oxon OX14 4RN

and by Routledge
605 Third Avenue, New York, NY 10158

Routledge is an imprint of the Taylor & Francis Group, an informa business

British Library Cataloguing-in-Publication Data
A catalogue record for this book is available from the British Library

Library of Congress Cataloging-in-Publication Data
A catalog record for this book has been requested

ISBN: 978-1-032-07979-0 (hbk)
ISBN: 978-1-032-07327-9 (pbk)
ISBN: 978-1-003-21235-5 (ebk)

DOI: 10.1201/9781003212355

Typeset in Goudy
by Apex CoVantage, LLC

Contents

Figures

Tables

Foreword (1)

The level of construction volume in Hong Kong has been exceptionally high over the past few years. Moreover, policy addresses in recent years show that the Hong Kong government is set to initiate a significant number of housing and infrastructure projects. As the construction market is expected to be prosperous in the coming years, both the public and private sectors are anticipated to grow substantially, offering many job opportunities to construction professionals, including quantity surveyors.

Quantity surveying is a key profession in the construction industry, and quantity surveyors in Hong Kong may practise in consulting firms or work with contractors, sub-contractors, developers or government departments. Irrespective of the employing organisation, a quantity surveyor is expected to provide core quantity surveying services that cover all aspects of procurement, contract management and project management. This book, *Quantity Surveying Practice: The Nuts and Bolts*, aims to support the critical role that quantity surveyors play in the expanding construction industry.

The general approach adopted in this book is to focus on the professional knowledge that is considered the bedrock of the skills and techniques essential for professional quantity surveyors. The topics chosen are those typically encountered by a quantity surveyor in all stages of a construction project, from inception to final accounting.

This book outlines the practical skills and contract management techniques needed by practising quantity surveyors, probationers and quantity surveying students. The fundamental concepts are also beneficial for new practitioners in the construction industry who have no formal training but are interested in pursuing the quantity surveying profession.

I believe that this book will serve as a practical and comprehensive guide for readers seeking to hone their professional skills in quantity surveying.

Sr Paul Wong Kwok Leung
Vice-President of The Hong Kong Institute of Surveyors
Past Chairman of The Hong Kong Institute of Surveyors
(Quantity Surveying Division)

Foreword (2)

As I approached the completion of my secondary school education, my father's experiences of the Great Depression of the 1930s and subsequent experience with a series of short-term jobs led him to impress on me the crucial need to find a steady job. As I had demonstrated little in the way of manual dexterity, he thought I might become an electrician or perhaps a draughtsman for a local manufacturing company in my hometown of Lancaster, UK. However, as fate would have it, he noticed an advertisement in the newspaper for a Junior Quantity Surveyor that, following some enquiries with some acquaintances, seemed to offer the possibility not only of a steady job, but a *profession*. I reluctantly applied – my love of Sherlock Holmes stories had led me to develop hopes (unsupported by my mother) of a possible career on the police force – and a short time later I found myself being interviewed by my future employer for a position that I had no knowledge about! However, as the sole applicant and having a good reference from my school (as well as saying that I was not afraid of hard work), I was taken on with the condition that I acquired the five O-level GCE subjects needed to become a member of the Royal Institution of Chartered Surveyors (RICS). I thus began my first full-time job with no idea what quantity surveying was about, and I quickly found myself checking calculations, greeting reps at the hole in the wall we used as a counter, and making morning tea and afternoon coffee in the one-man consulting practice to which I now belonged. Shortly thereafter, I obtained the five subjects needed, joined the RICS and embarked on part-time studies to learn what quantity surveying was all about and to answer the question 'What do quantity surveyors actually do?'

Surveyors are a cautious breed. Land surveyors, for example, traditionally measure a single angle eight times with their theodolites, each from a different approach – from the left, then the right, swing over 180 degrees and then repeat – to counter any weakness in the instrument. Quantity surveyors (QS) are no exception. Everything that can be checked is checked – every calculation, even transfers from one document to another. This is very much reflected in QS practices and procedures, as a QS seldom ventures into an activity with support. In construction work, measurements are carried out according to authorised rules (e.g., The Standard Method of Measurement of Building Works), which are published and adopted in each country (especially in former UK Commonwealth countries) and vary according to local traditions and conditions. Decisions about contractors' and clients' rights and responsibilities are dictated by authorised written contracts (e.g., Standard Form of Contract), leaving little room for discretion, at least in principle. This is the QS's *modus operandi*.

Surveyors' typical MO focuses on *money*. I have heard (usually from QSs themselves) that QSs tend to do what architects and engineers don't like doing. Architectural education and research are invariably concerned with *benefits* rather than *costs*. Quite understandably, since the distant pre-construction contractor days, when architects were responsible for the design *and* construction of buildings, the greatest function they provide (and where their greatest talent lies) is design and, to some extent, helping ensure that their design is realised. It is therefore unsurprising that architects tend to view QSs as inhibitory, seemingly bent on disallowing even the smallest luxury from being incorporated into an innovative design intended, by definition, to be out of the ordinary (as 'ordinary' is often associated with cheap in building work). The reality, as always, is that the QS's main task is to look after the client's financial needs. Staying within budget is a QS's primary goal and their *raison d'être*. The need to pay a consultant for this task is a big surprise to most inexperienced clients, who quite understandably see this as the responsibility of the architect. However, that is another story, one that is likely tied to the way the QS fees are calculated (students, please discuss!).

In my experience, civil engineers are more alert to the financial issues in construction work, and they may see its importance although not necessarily its complexity beyond the level of infrastructure. QSs are typically commissioned for large building projects, which can involve many sub-contractors and suppliers who do not necessarily work in harmony; hence, potential claims and conflicts can be both large and many, and their management is often the most onerous aspect of a QS's job. We therefore occasionally hear of civil engineers decrying the need for QSs – a site engineer once remarked to me that he could do the QS job on his lunch break!

Mechanical and electrical (M&E) engineers are, in my experience, 'a law unto themselves'. For large, specialised projects, such as hospitals, where M&E services can account for around 40% of the total building cost, their designs can go unchecked in the absence of a specialised M&E QS.

Calvin, Daisy and Sai On's work in producing this book very much reflects the QS practitioners' situation, practice and procedures. Unlike the UK's *Willis's Practice and Procedure for the Quantity Surveyor*, which focusses much more on current changes, *Nuts and Bolts*, as the title suggests, provides an excellent introduction to traditional practice in Hong Kong. The early chapters provide insight into the most fundamental estimating and tendering tasks of QSs in the pre-contract stages, as well as their post-contract responsibilities for interim valuation, valuing variations and preparation of final accounts. In addition, there are the more complex activities involved in claims and disputes – of which Professor Cheung is an acknowledged world authority. This structure accurately reflects my real-world experience as a consultant QS, where the usual day-to-day work revolved around my post-contract responsibilities in my personal portfolio of construction projects, which was interrupted from time to time by a team effort in the time-intensive procedure for preparing the tendering for a new project.

In reading through the chapters of this book, what caught my attention most is the level of rigour involved in QS practice and procedures in Hong Kong. Nothing is left to chance, it seems. There is a procedure for every usual – and unusual – situation. I doubt that this is the case outside Hong Kong, which makes the skills of its practitioners and students not only transferable but of considerable market value. As an educator, I was also very impressed with the book's format and the examples provided, especially in the estimation chapter, where the outcome was given first and the back-up calculations afterwards. Too often, I think, we tend to teach the build-up first, leaving the final product until later, but this often confuses students, who wind up not being able to see the wood for the trees. This is probably due to the nature of the QS role in the construction process, which, as I was well aware in my own early days in practice, provides an integral and vital contribution unlike any other profession.

Professor Martin Skitmore

Professor Adjunct of Construction Economics & Management
School of Built Environment, Queensland University of Technology, Australia

Foreword (3)

I have known the authors, Dr Calvin Keung, Dr Daisy Yeung and Dr Sai On Cheung, for years, if not for decades. I have also participated in many activities organised by the City University of Hong Kong, such as serving as their external academic advisor and as a juror of their Integrated Building Project Development coursework. Having had these opportunities to work closely with the authors, I have seen their passion and vision to uphold and to better the standard of the quantity surveying profession. They possess both the academic experience and the practical skills necessary to connect the market with the academic curriculum. I am so glad that they have written this book, which lays out the essential quantity surveying practices for anyone who is interested in the profession.

The book will be of great value to both students and practitioners. The book includes a general introduction, case studies and discussions that make it interesting to read and easy to digest. I am sure that it will also help experienced practitioners to refresh their basic QS techniques. This book is like a ship's anchor: it keeps you 'in position' irrespective of the ever-changing project requirements. These are the basic skills that a quantity surveyor must possess regardless of what tools are used, whether measurement software, BIM models or other digital applications.

Finally, I offer my congratulations to Dr Keung, Dr Yeung and Dr Cheung for successfully consolidating the enormous tasks of quantity surveying into a book that is concise and fundamental, as suggested by its name – *Quantity Surveying Practice: The Nuts and Bolts*. It may seem easy, but it is actually a tremendous accomplishment.

Mr Kenneth Kwan
Chairman, Rider Levett Bucknall

Preface

A quantity surveyor is sometimes described as an accountant or a lawyer of construction projects. Equipped with construction technology and engineering know-how, the quantity surveyor is specially trained to handle the costs and contractual matters of construction developments. Because of increasing expectations of quality and performance from project clients, the quantity surveyor must improve his professional skills and be able to solve the many intricate problems and disputes that arise in the demanding construction market. Thus, a practical guide is vital to facilitate the acquisition of professional knowledge and advance the profession of quantity surveying.

This book provides a practical guide to quantity surveying practice in the construction industry. Students learning quantity surveying in higher education institutions and probationers preparing for the Assessment of Professional Competence (APC) should consider this book as a reference guide because it focuses on the basic concepts underlying the technical aspects of quantity surveying. The market still lacks sufficient textbooks on local quantity surveying practice at both the pre- and post-contract stages. This book can also serve as a handy reference for young practicing surveyors because it specifically addresses common cost and contractual issues encountered in local building projects. Most importantly, this book contains many worked examples together with useful figures and real-life cases to help readers digest and understand the essentials of quantity surveying practice.

This book is organised and structured into the following seven chapters. Chapter 1 discusses the estimation of construction costs. Construction cost estimates are generally prepared by the quantity surveyor, and they evolve throughout the project lifecycle, becoming more detailed as more design information and details become available. This chapter introduces various estimating methods according to the working stages specified in the RIBA (Royal Institute of British Architects) Plan of Work 2020. To provide a comprehensive overview of construction cost estimates, the definition of floor areas, the sources of cost data and the general exclusions are also discussed. This chapter concludes with a discussion of the critical factors that influence estimation accuracy.

Chapter 2 gives an overview of tendering and tender documentation. Tendering is commonly used in the construction industry, as contractors are invited to bid for a construction project. The quantity surveyor plays a critical role in this process by preparing the tender documents. This chapter explains how contractors are invited to submit construction tenders and how the quantity surveyor is involved in the tender preparation process. It then provides a detailed explanation of the composition of a set of tender documents. Finally, it presents the concepts and functions of tender addenda and pre-tender estimates.

Chapter 3 examines the tender examination procedure and the approach to awarding the contract. The quantity surveyor is responsible for evaluating the returned tenders and providing professional advice to the employer regarding tender award. This chapter first introduces the tender analysis process, which includes tender opening and examination of tenders, including BQ rates and prices, and addresses the handling of tender queries and tender interviews. It then discusses the purposes and preparation of a tender report and summarises the typical content of such a report. Additionally, this chapter covers how the employer awards a tender to the successful tenderer and the essential documentation for contract execution.

Chapter 4 reviews the interim valuation process, from the contractor's submission of a payment application to the issuance of an interim valuation by the quantity surveyor and identifies the key issues within the process. An interim valuation is prepared by the quantity surveyor, who determines the payment amount stated as due to the contractor in an interim certificate. This chapter deals with the contract clauses for interim valuations and certificates. The prescribed payment time frame and mechanism are then discussed in detail. This chapter also provides some good valuation practices for the quantity surveyor preparing interim valuations.

Chapter 5 examines construction claims and discusses the key issues involved in establishing the contractual claims and the generic principles involved in their evaluation. To provide a foundation of knowledge, this chapter explains the basic concepts of claims arising from construction projects and the types of contractual claims that are legitimate under a building contract. This chapter also provides a brief analysis of some of the relevant contract provisions and addresses the important techniques used by the quantity surveyor to assess the contractual claims. Some illustrative examples of claim assessment are provided.

Chapter 6 addresses cost control and monitoring in construction projects. As the custodian of the employer's budget, the quantity surveyor should ensure that the authorised project budget is not exceeded. This chapter discusses the cost control process and examines the role of the quantity surveyor. This chapter also explains the key issues involved in cost monitoring with reference the project financial statements and financial reports prepared by the quantity surveyor. A detailed worked example of a project financial statement is illustrated at the end of the chapter for easy reference.

Chapter 7 discusses dispute management and three commonly used dispute resolution mechanisms: mediation, adjudication and arbitration. Two recent dispute resolution procedures that emphasise dispute avoidance are examined in detail: the dispute avoidance and resolution procedures used in the 2013 Hong Kong Housing Authority General Conditions for Building Works and the 2019 dispute resolution advisor system detailed in the Special Conditions of Contract Clause 59 of the Hong Kong Government General Conditions of Contract for Building Works. This chapter concludes with a discussion of design considerations for dispute resolution procedure through the use of a hypothetical project.

We hope that readers find this book helpful for understanding and learning quantity surveying practice. The authors have solid backgrounds and experience in the academic, practical and professional arenas, and we hope this book will become essential reading for students in surveying and construction management programmes and for APC probationers pursuing the professional quantity surveying pathway. It is also intended as a reference for practicing surveyors, and we recommend that it should be supplemented with any specific practice manuals or quality assurance procedures within the quantity surveyor's own office, together with a thorough knowledge of quantity surveying.

Keung Chung Wai Calvin PhD, FHKIS, FRICS

Yeung Kam Lan Daisy PhD, MHKIS, FRICS

Cheung Sai On DSc, PhD, FHKIS, FRICS

Acknowledgements

This book was fully supported by a sponsorship from the Hong Kong Institute of Surveyors. The authors greatly appreciate the extensive assistance, support, time and efforts of many parties, especially the Department of Architecture and Civil Engineering of the City University of Hong Kong, as well as the colleagues and editorial team who worked on this book.

Notes to readers

Unless otherwise clarified in the individual chapters.

1   The term 'the quantity surveyor' refers to the quantity surveyor serving as the construction cost consultant.
2   The book is heavily based on the *Agreement & Schedule of Conditions of Building Contract* for use in the Hong Kong Special Administrative Region, which is the 2005 edition of the *Standard Form of Building Contract* published by the Hong Kong Institute of Architects, the Hong Kong Institute of Construction Managers and the Hong Kong Institute of Surveyors.

1 Estimation

Learning outcomes

Upon completion of this chapter, you should be able to do the following.

1 Understand how an architectural work plan relates to different estimation methods.
2 Be able to calculate gross floor area (GFA) and construction floor area (CFA).
3 Use relevant cost data to make estimates.
4 Understand the various estimation methods used in different work stages.

1.1 Introduction

What is the first question asked by an employer who wants to construct a building? In most cases, it is, 'How much will it cost?'. This chapter discusses the purpose of cost estimation and various estimation methods. The critical factors affecting estimation accuracy are also explored, and a case study is provided at the end of the chapter.

1.2 Architectural work plan

What is a plan of work? The RIBA Plan of Work 2020 published by the Royal Institute of British Architects (RIBA) defines a work plan as follows:

> Many countries have no formal process for designing a building. 'The way to do it' is unwritten and unrecorded, with informal processes handed down from one generation of professionals to the next. Regardless of where in the world a building is erected, the core tasks are the same.
>
> 1 Agree on appointments with the professional team.
> 2 Develop a brief with the employer.
> 3 Create design concept options.
> 4 Coordinate design development.
> 5 Prepare a planning application.
> 6 Apply for planning consent.
> 7 Assemble construction information.
> 8 Prepare a tender.
> 9 Obtain consent required prior to construction.
> 10 Obtain a building contract.

DOI: 10.1201/9781003212355-1

11 Construct the building.
12 Inspect the construction as it progresses.
13 Hand over the building.

Based on the core tasks above, work can be categorised into the following stages.

1 Pre-design stage
2 Design stage
3 Construction stage
4 Handover stage
5 In-use stage
6 End-of-life stage

Table 1.1 illustrates the components of the pre-design and design stages of six international plans of work.

The estimation method used depends on the following factors.

Table 1.1 Comparison of international plans of work

	Pre-Design		Design			
RIBA (UK)	Strategic Definition	Preparation and Brief	Concept Design	–	Developed Design	Technical Design
ACE (Europe)	Initiative	Initiation	Concept Design	Preliminary Design	Developed Design	Detail Design
AIA (USA)	–	–	Schematic Design	–	Design Development	Construction Document
NZCIC (NZ)	–	Pre-design	Concept Design	Preliminary Design	Developed Design	Detail Design
HKIA (HK)	*No specific stages are defined, but in general, the process consists of the following stages.*					
	Strategic Definition	Preparation and Brief	Concept Design	Preliminary Design	Developed Design	Detail Design

Source: RIBA (2020)

1 The stage of development (refer to Table 1.1).
2 The project information available at the work stage.
3 The time available for preparing the estimate.
4 The estimator's preference for and familiarity with the methods adopted.
5 The estimator's experience.

In the following sections, estimation methods are discussed for the following work stages (see Table 1.1).

1 Strategic Definition (Feasibility study stage)
2 Preparation and Brief (Brief design stage)
3 Concept/Preliminary Design (Preliminary conceptual design)
4 Developed Design (Design development from scheme design up to working drawings)
5 Detail Design (Completion of working drawings)

1.3 Area definition for estimation and cost planning

1.3.1 Gross floor area (under lease)

Definition of gross floor area (under lease) in Building (Planning) Regulations 23(3)(a):

The area contained within the external walls of the building measured at each floor level (including any floor below the level of the ground), together with the area of each balcony in the building, which shall be calculated from the overall dimensions of the balcony (including the thickness of the sides thereof), and the thickness of the external walls of the building.

Exemptions

According to the Practice Notes for Authorized Persons (PNAP) and Joint Practice Notes (JPN), the following areas are not included in the calculation of gross floor area (Building Department, 2011).

1   Mechanical rooms
2   Refuge floors
3   Hotel back-of-house areas
4   Lift shafts
5   Recreational areas in residential developments
6   Balcony/utility platforms
7   Precast wall panels
8   Space occupied by curtain walls

1.4 Construction floor area (CFA) and gross covered floor area

Arcadis (2020a), a Hong Kong construction cost consultant, gives the following definitions of construction floor area (CFA).

1   Covered floor areas fulfilling the functional requirements of the building measured to the outside face of the external walls (or, in the absence of such walls, the external perimeter) of the building.
2   Floor areas occupied by walls, columns, stairwells, lift shafts, plant rooms, water tanks, balconies, utilities platforms, vertical ducts, service floors higher than 2.20 m and the like.
3   CFA excludes bay windows (although they may be included under different definitions from other cost consultants or institutions), air-conditioner hoods, architectural fins, planters projected from the building and areas covered by canopies, roof eaves and awnings.

1.4.1 Cost data for estimation

Data sources for preparing an estimate include the following (Table 1.2 presents the cost data abstracted from a Hong Kong construction cost consultant).

1   Historical data from the quantity surveyor's office.
2   Online data sources (e.g., construction cost books from local cost consultant firms).
3   Quotations from specialists and suppliers.

Table 1.2 Construction costs for Hong Kong

CONSTRUCTION COSTS FOR HONG KONG			
Building Type	*Building (HK$)*	*Services (HK$)*	*Total (HK$)*
Domestic			
Apartment, high-rise, public authority standard	9,050 to 10,950	1,950 to 2,350	11,000 to 13,500
Apartment, high-rise, average standard	19,700 to 21,700	3,900 to 5,600	23,600 to 27,300
Apartment, high-rise, high-end	xxxx to xxxx	xxxx to xxxx	xxxx to xxxx
Terraced houses, average standard	xxxx to xxxx	xxxx to xxxx	xxxx to xxxx
Detached houses, high-end	xxxx to xxxx	xxxx to xxxx	xxxx to xxxx
Office/Commercial			
Medium/high-rise offices, average standard	xxxx to xxxx	xxxx to xxxx	xxxx to xxxx
High-rise offices, prestige quality	xxxx to xxxx	xxxx to xxxx	xxxx to xxxx
Out-of-town shopping centre, average standard	xxxx to xxxx	xxxx to xxxx	xxxx to xxxx
Hotels			
Budget hotels – 3-star, mid-market	22,750 to 23,150	6,950 to 8,450	29,700 to 31,600
Business hotels – 4/5-star	xxxx to xxxx	xxxx to xxxx	xxxx to xxxx
Luxury hotels – 5-star	xxxx to xxxx	xxxx to xxxx	xxxx to xxxx
Industrial			
Owner-operated factories, low-rise, lightweight industry	xxxx to xxxx	xxxx to xxxx	xxxx to xxxx

Source: Arcadis (2020b)

Remarks:

1   Because of issues related to intellectual property rights, this table is for demonstration purposes only. Some of the cost data have thus been replaced with 'xxxx'.
2   Only the unit rates (cost range) used in Example 1 are shown and are highlighted, as in the latter part of this chapter.

1.5 Estimation methods adopted at various work stages

1.5.1 Feasibility study stage

The information available at this stage includes (i) gross floor area, (ii) location, (iii) preliminary site information and (iv) the required standards. Drawings are typically not included at this stage. Early cost estimates are prepared for the employer only.

The estimated cost is calculated as construction unit cost/rate x area or unit. Some examples are as follows:

1   Cost per m^2 (for residential, commercial or office development)
2   Cost per bed (for hospitals)
3   Cost per room (for hotels)
4   Cost per seat (for auditoriums and cinemas)
5   Cost per space (for carparks)

Table 1.3 Example 1: Estimation at the feasibility study stage

Function	Area or Unit	Construction Unit Cost/ Rate	Calculations	Total Indicated Cost HK$
Residential (assume ordinary standard)	20,000 m²	#HK$24,000 per m²	20,000 m² (Area) × HK$24,000 per m² (Unit rate)	480 Million
Hospital	500 beds	HK$2,500,000 per bed	500 bed × HK$2,500,000 per bed	1,250 Million
Hotel (3-star standard)	100 rooms	#HK$30,000 per room	100 rooms × HK$30,000 per room	3 Million
Car Park	300 spaces	HK$16,000 per space	300 no. × HK$16,000 per space	4.8 Million

Remarks:

1   # Cost data range is from Table 1.2.
2   Other cost data are hypothetical and are for education purposes only.

1.5.2 Brief design stage

The information available at this stage includes (i) outline schematic drawings showing the shape and size of the building with minimal layout design and (ii) the standards required. In this stage, a preliminary estimation of elements is conducted. The individual elements (e.g., carcase, façade, finishing, etc.) are priced using data from cost analyses of similar projects. Elemental cost data should be adjusted based on (i) quality, (ii) quantity and (iii) current price level. This estimate usually forms the basis of a preliminary budget.

[See Table 1.4: Example 2: Brief design stage – Preliminary elemental estimate in the latter part of this chapter]

1.5.3 Preliminary conceptual design stage

The information available at this stage includes (i) preliminary drawings with plans, elevations and sections showing the building shape, internal layout, windows, doors, etc., (ii) structural framing plans, (iii) foundation information, (iv) standards of finishes and (v) building service work requirements. At this stage, detailed estimates of elements are made based on the quantities approximated at current prices and are supplemented by data from cost analyses for items that cannot be quantified at this stage with the price level adjusted.

[See Table 1.5: Example 3: The cost plan for completion of the conceptual design in the latter part of this chapter]

When using historical elemental cost analyses of other similar projects, three major areas should be noted, and necessary adjustments should be made.

1   Quantity

Quantity adjustment depends on the different quantities of elements (adjustments made to elemental quantity; e.g., increase reinforcement by 30%).

2   Quality

Quality adjustment must be made with reference to the available specifications and design drawings (adjustments to the elemental unit rate; e.g., change the kitchen floor tile from ceramic to marble tile, change site conditions, change contract conditions and project characteristics, etc.).

3   Price

Price adjustment must be made based on the tender date (adjustment of different price levels; e.g., adjust the tender price index from the 1st quarter of 2019 to the 4th quarter of 2020).

1.5.4 Developed design (design development from the scheme design stage up to the working drawings stage)

More design details become available as the design is developed in the working drawing stage. Different elemental estimates are prepared at different design stages – these estimates are to be reconciled with the budget, or the budget is to be revised according to the updated estimate. If there is any major change, the employer and the architect should be informed. A cost-saving exercise or alternative design study may be required if the estimated cost exceeds the previous budget.

Alternative design estimate

In this stage, the architect may propose alternative designs to the employer's consideration and approval. Comparative cost studies are required.
   [See Table 1.6: *Example 4: The cost comparison of designs for phase 1 and 2 in the latter part of this chapter*]
   The cost comparison studies that may be conducted at various stages are tabulated in Table 1.7.

Example 4: The cost plan for completion of conceptual design for phase 2

Table 1.7 Cost comparison studies at various stages

Stage	Types of Design Work Involved	Examples of Cost Comparison Studies	Examples
Feasibility Study Stage	Preliminary Design Scheme	Type of building	Hotel vs office
		Alternative sites	Location of hospital at Shatin vs Hong Kong Island
		Alternative standards	Grade A vs grade B office
		Phasing options	1 phase vs 2 or 3 phases

Stage	Types of Design Work Involved	Examples of Cost Comparison Studies	Examples
Outline Conceptual Design Stage	Outline Design Development Options	No. of storeys	25 storeys with large footprint vs 50 storeys with small footprint
		Alternative plan shape	Rectangular vs circular
		Basement provision	Basement carpark vs podium carpark
		Form of construction	Steel frame vs reinforced concrete frame
		Future expansion	Provision of vertical expansion vs horizontal expansion
Design Development Stage Up To Working Drawings	Detailed Design Alternatives	Alternative materials	Carpet vs timber flooring
		Alternative elevation studies	Curtain wall vs aluminum window
		Alternative mechanical & electrical systems	Central air conditioning vs split air conditioning
		Alternative internal layouts	6 flats per floor vs 8 flats per floor for residential development
		Alternative provisions	Swimming pool vs tennis court for external works
		Building storey height	2.80 m vs 3.20 m storey height for residential development

Remarks:

See Table 1.6: Example 4: Cost comparison of designs in the former part of this chapter.

Important steps and methods in an estimation conducted at the detailed design stage

Preparation of estimates for the detailed design stage include the following steps.

1   Estimate checklist.
2   Measure CFA, perimeter and cost-significant items with a certain degree of accuracy.
3   Ignore minor details.
4   Include composite items and rates, e.g., tiling + screed (measured in overall m²), door + doorframe + paint + ironmongery (e.g., door hinge, handle, etc.) (measured in no. overall).
5   Include lump sum allowances, e.g., preliminaries, contingencies etc.
6   Obtain rates from historical cost data/cost analysis, recent tenders returned, recent estimates prepared and quotations received.

PREPARATION OF A DETAILED DESIGN ESTIMATE

1. Measure quantities (q)

2. Search for cost data/cost analyses of similar projects

3. Extract appropriate rates (r)

4. Adjust rates for quantity, time and quality differences

5. Make allowances for items that are not included in the above steps

6. Calculate initial estimates ($\sum qr$) with appropriate adjustments

7. Add preliminaries %

8. Add contingencies %

9. Sum up steps 6 to 8

Figure 1.1 Method for preparing a detailed design estimate

1.5.5 *Technical design development stage (completion of working drawings)*

The information at this stage is well developed and can be used by the quantity surveyor to prepare bills of quantities (based on a full set of detailed working drawings). At this stage, a pre-tender estimate can be made by pricing the measured bills of quantities; this estimate will be used as a reference during tendering and for evaluating the tenders.

1.6 General exclusions from the estimate

The following items should be excluded from construction cost estimates, and a list of these exclusions should be included in the estimate.

1 Land cost, conversion premium.
2 Financing charges and developer's overhead.
3 Professional and legal fees.
4 Advertising, promotion and sales agent fees.
5 Provision of adequate utility services and drainage facilities in the vicinity of the development site.
6 Fitting out works to tenant areas (e.g., for office or retail projects) or other areas.
7 Operational items, equipment and machinery provided by the employer (e.g., for hotel or factory projects).
8 Works outside the scope of the project (detailed elaboration is recommended).
9 Fluctuations in construction costs from the date of estimate to the date of tender.

The above exclusion list is not exhaustive and is for reference only. It may be changed subject to the requirements of the project.

1.7 Information included in the estimate

The following information should be provided in the estimate.

1 Summary of estimated cost (see Example 2- Table 1.4).
2 Exclusion list.
3 Elemental breakdowns (see Example 3- Table 1.5).
4 Scope of works (i.e., brief project description).
5 Area schedule (i.e., GFA or CFA).
6 List of drawings and sketches.
7 Outline specification or finishing schedule (in table format).
8 Breakdown of estimate (i.e., breakdown of approximate quantities and allowances).
9 Assumptions made in the estimate, if any.
10 The architect's or engineer's design advice for preparation of the estimate, if any.
11 Supplementary information (e.g., anticipated cash flow forecast).

1.8 Accuracy of early cost forecasts

Early estimation of construction costs should be regarded as an indicator of cost allowance rather than a firm and precise cost. The accuracy will increase as the design is developed. Critical factors affecting estimation accuracy are as follows.

1 Different estimation techniques have different weaknesses.
2 Project variance (e.g., design, site location, procurement and contractual arrangement, construction method, etc.).

3   Economic environment (e.g., market price, exchange rates, etc.).
4   Availability of information (e.g., design information, cost data, etc.).
5   Experience of the quantity surveyor.

1.9 Chapter summary

This chapter reviewed the estimation methods used at different design stages. It introduced (1) cost indication at the very early stage and (2) elemental cost estimate at the preliminary design, schematic and final design stages, and provided examples. Five critical factors to enhance the accuracy of early cost forecasts were presented. The quantity surveyor is advised to take these factors into account in cost estimations.

1.10 Case study

1.10.1 *Background information for the hypothetical project*

The proposed residential development site is divided into two phases. Phase 1 consists of five houses with 3.00 m storey height for all floors, with construction commencing in the first phase. Phase 2 consists of five houses with 5.00 m storey height at G/F and 3.00 m storey height at 1/F, with construction commencing in the second phase. All drawings for the case are indicated in Figure 1.2 to 1.10.

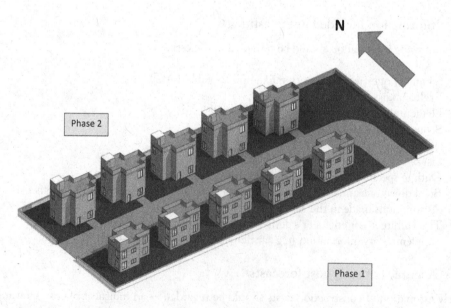

Figure 1.2 Perspective view of the hypothetical residential development project

Drawings for Phase 1

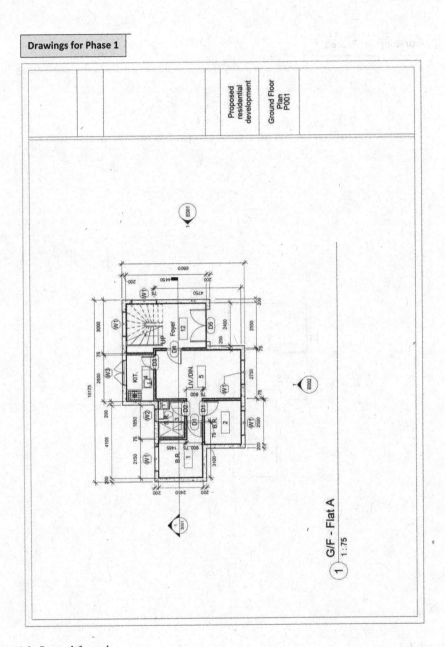

Figure 1.3 Ground floor plan

Drawings for Phase 1

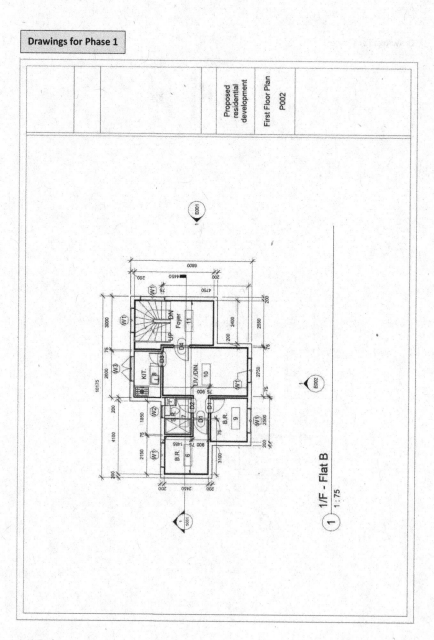

Figure 1.4 1/F plan

Drawings for Phase 1

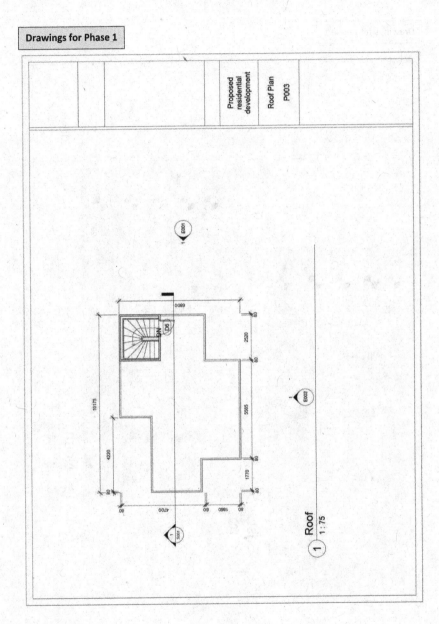

Figure 1.5 Roof plan

Drawings for Phase 1

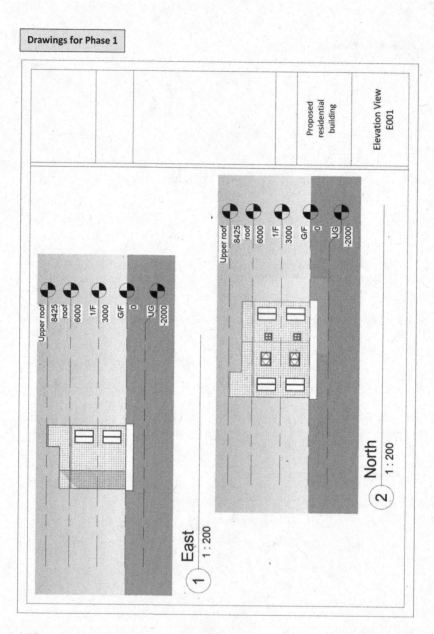

Figure 1.6 Elevation view E001

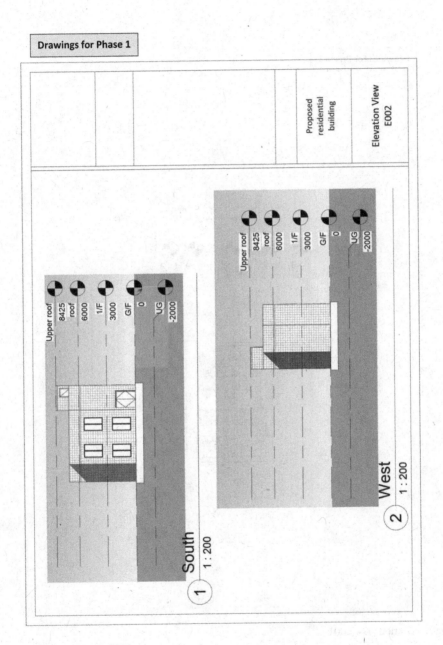

Figure 1.7 Elevation view E002

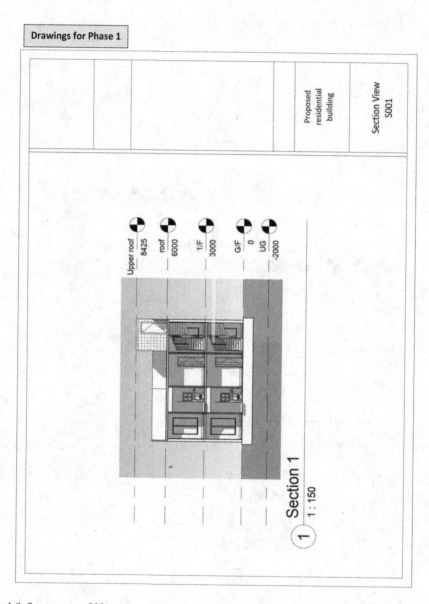

Figure 1.8 Section view S001

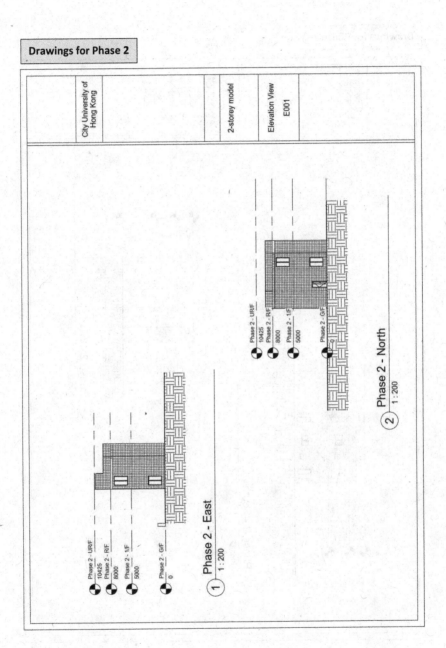

Figure 1.9 Elevation view E001

Drawings for Phase 2

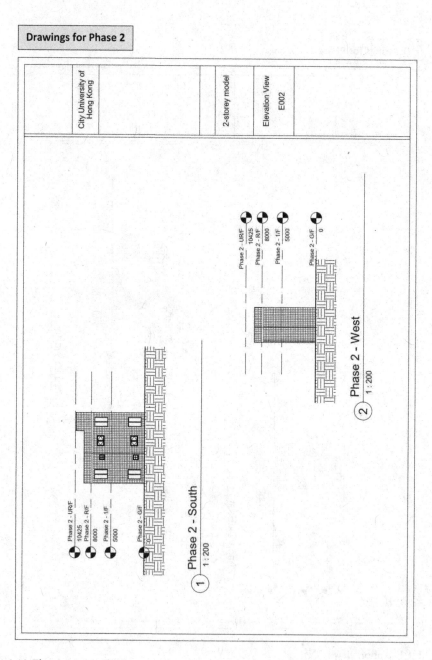

Figure 1.10 Elevation view E002

1.10.2 Examples for estimation method adopted at various work stages

Example 2: Brief design stage – preliminary elemental form of estimate for phase 1

Table 1.4 Example 2: Brief design stage – preliminary elemental form of estimate for phase 1

(a) The C.F.A Calculation for 1 House

Level	Area (CFA)
G/F	52 m2
1/F	52 m2
R/F	6 m2
Total	110 m2

Note:
Example 2 (where only preliminary design is available)

It is a preliminary cost plan which is calculated by C.F.A × HK$/m2 (elemental unit cost of similar project)

(b) Brief Elemental Cost Plan (at very preliminary stage)

Summary of Estimate (Preliminary Estimate)

Total Site Area:	3,620 m2
Total Gross Floor Area (GFA):	483 m2 (5,195 sq.ft)
Total Construction Floor Area (CFA):	550 m2 = 110 m2 CFA × 5 houses
CFA/GFA ratio:	1.14

	CFA (m2)	*Phase 1* Construction Cost (HK$)	Unit Cost (HK$/m2 CFA)	(HK$/sq.ft GFA)
1 Site Investigation		605,000*	1,100	116
2 Hoarding		2,035,000*	3,700	392
3 Site Formation Works		2,200,000*	4,000	423
4 Foundation and Substructure (pending engineer's input, assumed raft foundation for houses)		1,100,000*	2,000	212
5 Superstructure	550	13,750,000	25,000	2,647
5.1 House A – 5 nos.	550	13,750,000*	25,000	2,647
6 External Works and Landscaping		26,235,000	47,700	5,050
6.1 External landscaping, paving and EVA		18,150,000*	33,000	3,494
6.2 Paving/pedestrian walkway outside site boundary		1,100,000*	2,000	212
6.3 Utilities within the site		6,050,000*	11,000	1,165
6.4 Underground drainage		605,000*	1,100	116
6.5 Utilities connections		330,000*	600	64
Sub-total	550	45,925,000	83,500	8,840
7 Preliminaries (15% of Item 1–6) rounded up to		6,900,000	12,546	1,328
8 Contingencies (10% of Item 1–7) rounded up to		5,300,000	9,636	1,020
9 Fluctuation		1,500,000	2,727	289
TOTAL CONSTRUCTION COST (at January 2020 Price Level)		59,625,000	108,409	11,477

Remark:-
* The costs are allowed figures for use in the presentation of the estimating exercise only and no further breakdown has been included.

Example 3: The cost plant for the completion of conceptual design for phase 1

Table 1.5 Example 3: The cost plan for completion of conceptual design for phase 1

(a) The C.F.A. Calculation for 1 House

Level	Area (CFA)
G/F	52 m2
1/F	52 m2
R/F	6 m2
Total	110 m2

> **Note:**
> Example 3 (where conceptual design is developed)
> It is estimated from conceptual design information, and it involves the combination of estimate method as (a) approximate quantities x current complete unit rates
> (where measured quantities are available);
>
> (b) C.F.A x HK$/m2 (elemental unit cost of similar project)
> (where measured quantities are not available due to insufficient design information)

(b) Elemental cost plan

Summary of Estimate (Preliminary Estimate)

Total Site Area:	3,620 m2
Total Gross Floor Area (GFA):	483 m2 (5,195 sq.ft.)
Total Construction Floor Area (CFA):	550 m2 =110 m2 CFA × 5 houses
CFA/GFA ratio:	1.14

	CFA (m2)	Construction Cost (HK$)	Phase 1 Unit Cost (HK$/m2 CFA)	(HK$/sq.ft GFA)
1 Site Investigation		600,000*	1,091	115
2 Hoarding		2,000,000	3,636	385
3 Site Formation Works		2,200,000	4,000	423
4 Foundation and Substructure		1,080,000	1,964	208
(pending engineer's input, assumed raft foundation for houses)				
5 Superstructure	550	13,280,000	24,145	2,556
5.1 House A – 5 nos.	550	13,280,000	24,145	2,556
6 External Works and Landscaping		23,550,000	42,818	4,533
6.1 External landscaping, paving and EVA		15,820,000	28,764	3,045
6.2 Paving/pedestrian walkway outside site boundary		1,100,000*	2,000	212
6.3 Utilities within the site		5,700,000*	10,364	1,097
6.4 Underground drainage		600,000*	1,091	115
6.5 Utilities connections		330,000*	600	64
Sub-total	550	42,710,000	77,655	8,221
7 Preliminaries (15% of Item 1–6)　rounded up to		6,500,000	11,818	1,251
8 Contingencies (10% of Item 1–7)　rounded up to		5,000,000	9,091	962
10 Fluctuation		1,455,000	2,645	280
TOTAL CONSTRUCTION COST (at January 2020 Price Level)		55,665,000	101,209	10,715

Remark:
* The costs are allowed figures for use in the presentation of the estimating exercise only and no further breakdown will be included.

Table 1.5 Example 3: The cost plan for completion of conceptual design for phase 1 (Cont'd)

Back-up Calculations

Descriptions	Quantity	Unit	Rate	Estimated Cost
			HK$	HK$
2 Hoarding	265	m	7,500	1,987,500
a Hoarding			Total	1,987,500
			Say	2,000,000
3 Site Formation Works				
a Excavation; assumed 1m depth	3,620	m3	250	905,000
b Cart away	3,620	m3	340	1,230,800
			Total	2,135,800
			Say	2,200,000
4 Foundation and Substructure				
Total building footprint:	260 m2			
Total building footprint with raft foundation:	260 m2			
a Raft foundation footing; assumed 1m depth thick	260	m2	4,000	1,040,000
b Blinding layer	260	m2	130	33,800
			Total	1,073,800
			Say	1,080,000
5 Superstructure				
Back-up calculations refer to next pages				
6.1 External Works and Landscaping				
Site Area:	3,620 m2			
Less: G/F CFA of Houses:	260 m2			
External landscaping, paving and EVA:	3,360 m2			
a External landscaping, paving and EVA including drainage, lighting, etc.	3,360	m2	4,000	13,440,000
b 2.5m high fence wall along site boundary; painted finish, design to budget	260	m	8,000	2,080,000
c Main entrance gate		sum		100,000
d Signage		sum		100,000
e Guard house	1	No.	100,000	100,000
			Total	15,820,000
			Say	15,820,000

Table 1.5 Example 3: The cost plan for completion of conceptual design for phase 1 (Cont'd)

Elemental Breakdown for 5. Superstructure

Total Construction Floor Area (CFA):	550 m2	= 110 m2 x CFA x 5houses
		Phase 1

Elements	Elemental Total	Elemental Unit Cost
	(HK$)	(HK$/m2)
1. Carcase		
1.1 Frame and slab	2,770,000	5,036
1.2 Internal walls	610,000	1,109
1.3 Doors and shutters	450,000	818
Sub-total for Item 1	**3,830,000**	**6,964**
2. Facade		
2.1 External walls	1,230,000	2,236
2.2 External wall finishes	520,000	945
2.3 Windows	690,000	1,255
Sub-total for Item 2	**2,440,000**	**4,436**
3. Finishings		
3.1 Roof finishes	70,000	127
3.2 Floor finishes	810,000	1,473
3.3 Internal wall finishes	590,000	1,073
3.4 Ceiling finishes	270,000	491
3.5 Decor, graphics and signage	730,000	1,327
Sub-total for Item 3	**2,470,000**	**4,491**
4. Furniture and Fittings		
4.1 Built-in furniture	1,060,000	1,927
4.2 Metal works and sundries	60,000	109
4.3 Artwork	500,000	909
4.4 Equipment	450,000	818
4.5 Special light fittings	10,000	18
Sub-total for Item 4	**2,080,000**	**3,782**
5. Building Services		
5.1 Sanitary fittings	200,000	364
5.2 Plumbing and drainage	550,000	1,000
5.3 Electrical	830,000	1,509
5.4 Fire services	280,000	509
5.5 Mechanical ventilation and air- conditioning	370,000	673
5.6 Gas	110,000	200
5.7 Builders work, profit and attendance (5%) of Item 5 Building Services	120,000	218
Sub-total for Item 5	**2,460,000**	**4,473**
Total for Superstructure (Phase 1)	**13,280,000**	**24,145**

Table 1.5 Example 3: The cost plan for completion of conceptual design for phase 1 (Cont'd)

Back-up Calculations for 5. Superstructure
Total Construction Floor Area (CFA): 550 m2 = 110 m2 × CFA × 5houses

Descriptions	Quantity	Unit	Phase 1	
			Rate	Estimated Cost
			(HK$)	(HK$)
1.1 Frame and Slabs				
Horizontal elements				
a. Reinforced concrete, grade 45D; to slab and beam	133	m3	1,300.00	172,900
b. Rebar to slab and beam (230kg/m3)	30,590	kg	12.00	367,080
c. Formwork to slab and beam	665	m2	350.00	232,750
d. Allow 5% for miscellaneous (rounded up)				40,000
			Total for Horizontal elements:	812,946
Vertical elements				
a. Reinforced concrete, grade 45D; to structural wall	227	m3	1,300.00	295,100
b. Rebar to structural wall (280kg/m3)	63,560	kg	12.00	762,720
c. Formwork to structural wall	2,268	m2	350.00	793,800
d. Allow 5% for miscellaneous (rounded up)				100,000
			Total for Vertical elements:	1,951,620
			Total	2,764,350
			say	2,770,000
1.2 Internal walls				
a. 75mm thick concrete block wall	546	m2	700.00	382,200
b. Glass partition for shower	45	m2	5,000.00	225,000
			Total	607,200
			Say	610,000
1.3 Doors and shutters				
a. Double leaf doors to flat entrances	5	no.	20,000.00	100,000
b. Single leaf to bedroom	20	no.	6,000.00	120,000
c. Single leaf doors to bathroom	10	no.	6,000.00	60,000
d. Single leaf doors to kitchen	10	no.	7,000.00	70,000
e. Single leaf doors to living/dining room	10	no.	6,000.00	60,000
f. Single leaf doors to roof	5	no.	7,000.00	35,000
			Total	445,000
			Say	450,000
2.1 External walls				
a. 200mm thick reinforced concrete wall	1,079	m2	1,100.00	1,186,900
b. 150mm thick reinforced concrete parapet wall	33	m2	1,100.00	36,300
			Total	1,223,200
			Say	1,230,000

Table 1.5 Example 3: The cost plan for completion of conceptual design for phase 1 (Cont'd)

Back-up Calculations for 5. Superstructure
Total Construction Floor Area (CFA):   550 m2   = 110 m2 × CFA × 5houses

Descriptions	Quantity	Unit	Rate	Phase 1 Estimated Cost
			(HK$)	(HK$)
2.2 External wall finishes				
a. Ceramic tiles (P.C. HK$50/m2) to external wall	1,079	m2	450.00	485,550
b. Ceramic tiles (P.C. HK$50/m2) to parapet wall	66	m2	450.00	29,700
			Total	515,250
			Say	520,000
2.3 Windows				
a. Window to bedroom	56	m2	3,500.00	196,000
b. Window to bathroom	6	m2	3,500.00	21,000
c. Window to kitchen	11	m2	3,500.00	38,500
d. Window to living/dining room	28	m2	3,500.00	98,000
e. Window to internal staircase	56	m2	6,000.00	336,000
			Total	689,500
			Say	690,000
3.1 Roof finishes				
a. Tiles including cement sand screed, insulation and waterproofing	46	m2	1,300.00	59,800
b. Waterproofing and insulation only	6	m2	500.00	3,000
c. Allow for skirting (10%)		sum		6,280
			Total	69,080
			Say	70,000
3.2 Floor finishes				
a. Natural stone to house entrance	33	m2	1,400.00	46,200
b. Stone finishes to bedroom	108	m2	1,400.00	151,200
c. Stone finishes to bathroom	29	m2	1,400.00	40,600
d. Stone finishes to kitchen	38	m2	1,400.00	53,200
e. Stone finishes to living/dining room	205	m2	1,400.00	287,000
f. Stone finishes to internal staircase; including nosing tiles	107	m2	1,400.00	149,800
g. Allow for skirting (10%)		sum		72,800
			Total	800,800
			Say	810,000

Table 1.5 Example 3: The cost plan for completion of conceptual design for phase 1 (Cont'd)

Back-up Calculations for 5. Superstructure
Total Construction Floor Area (CFA):   550 m2   = 110 m2 × CFA × 5houses

Descriptions	Quantity	Unit	Rate	Phase 1 Estimated Cost
			(HK$)	(HK$)
3.3 Internal wall finishes				
a. Plaster with emulsion paint to house entrance	147	m2	200.00	29,400
b. Plaster with emulsion paint to bedroom	592	m2	200.00	118,400
c. Glazed ceramic tiles to bathroom	207	m2	650.00	134,550
d. Glazed ceramic tiles to kitchen	240	m2	500.00	120,000
e. Plaster with emulsion paint to living/dining room	662	m2	200.00	132,400
f. Plaster with emulsion paint to internal staircase	266	m2	200.00	53,200
			Total	587,950
			Say	590,000
3.4 Ceiling finishes				
a. Plaster with emulsion paint to house entrance	33	m2	200.00	6,600
b. Plaster with emulsion paint to bedroom	108	m2	200.00	21,600
c. Gypsum board suspended ceiling to bathroom	29	m2	1,500.00	43,500
d. Gypsum board suspended ceiling to kitchen	38	m2	1,500.00	57,000
e. Plaster with emulsion paint to living/dining room	205	m2	200.00	41,000
f. Plaster with emulsion paint to internal staircase	107	m2	200.00	21,400
g. Allow for bulkhead	5	house	15,000.00	75,000
			Total	266,100
			Say	270,000
3.5 Decor, Graphics and Signage				
a. Decor, graphics and signage		sum		730,000
			Total	730,000
			Say	730,000
4.1 Built-in furniture				
a. Kitchen cabinet with worktop	10	no	90,000.00	900,000
b. Vanity counter with marble countertop and mirror cabinet to bathroom	10	no	15,000.00	150,000
c. Letter box	5	house	1,500.00	7,500
			Total	1,057,500
			Say	1,060,000
4.2 Metal works ad sundries				
a. Metal works and sundries	550	m2	100.00	55,000
			Total	55,000
			Say	60,000

Table 1.5 Example 3: The cost plan for completion of conceptual design for phase 1 (Cont'd)

Back-up Calculations for 5. Superstructure
Total Construction Floor Area (CFA):   550 m2   = 110 m2 × CFA × 5 houses

				Phase 1
Descriptions	*Quantity*	*Unit*	*Rate*	*Estimated Cost*
			(HK$)	(HK$)
4.3 Artwork				
a. Artwork at house entrance	5	no	100,000.00	500,000
			Total	500,000
			Say	500,000
4.4 Equipment				
a. Kitchen appliances	10	house	40,000.00	400,000
b. Electric water heater to kitchen	10	house	5,000.00	50,000
			Total	450,000
			Say	450,000
4.5 Special light fittings				
a. Special lighting fittings for house entrance		sum		10,000
			Total	10,000
			Say	10,000
5.1 Sanitary fittings				
a. Sanitary fittings to bathroom	10	no	15,000.00	150,000
b. Allow for bathroom accessories	10	set	5,000.00	50,000
			Total	200,000
			Say	200,000
5.2 Plumbing and drainage				
a. Plumbing and drainage	550	m2	1,000.00	550,000
			Total	550,000
			Say	550,000
5.3 Electrical				
a. Electrical	550	m2	1,500.00	825,000
			Total	825,000
			Say	830,000
5.4 Fire Services				
a. Fire services	550	m2	500.00	275,000
			Total	275,000
			Say	280,000

Table 1.5 Example 3: The cost plan for completion of conceptual design for phase 1 (Cont'd)

Back-up Calculations for 5. Superstructure
Total Construction Floor Area (CFA):   550 m2   = 110 m2 x CFA x 5houses

				Phase 1
Descriptions	Quantity	Unit	Rate	Estimated Cost
			(HK$)	(HK$)
5.5 Mechanical Ventilation and Air Conditioning				
a. Split type air conditioning to house entrance	5	no	20,000.00	100,000
b. Thermal ventilator to bedroom	20	no	6,000.00	120,000
c. Ventilation fans to bathroom and kitchen	20	no	6,000.00	120,000
d. Allow for bulkhead of mechanical ventilation	5	house	6,000.00	30,000
			Total	370,000
			Say	370,000
5.6 Gas				
a. Gas	550	m2	200.00	110,000
			Total	110,000
			Say	110,000
5.7 Builders works 5% of Building Services		Sum		117,000
			Total	117,000
			Say	120,000
		Total of 5. Superstructure (Phase 1)		13,280,000

Table 1.5 Example 3: The cost plan for completion of conceptual design for phase 1 (Cont'd)

FORECAST OF CONSTRUCTION COST FLUCTUATION
Price Level in Estimate: July 2018

* Assumed fluctuation percentage as follows:-
January 2018 to December 2018: +1%
January 2019 to December 2019: +3%
January 2020 to December 2020: -1%

Works Packages	Phase 1								Year 2018 +1%	Year 2019 +3%	Year 2020 -1%
	Anticipated Construction Sum as of July 2018 Level	Anticipated Contract Sum (Included Preliminaries and Contingencies)	Anticipated Cost of Inflation upto Tender Date	Total Construction Cost Including Inflation	Anticipated Tender Award Date	Percentage Calculation	Fluctuation Percentage	Fluctuation Period			
	(HK$)	(HK$)	(HK$)	(HK$)		Formula	(% p.a.)	(Month)			
1. Site Investigation	600,000	762,000	2,000	764,000	October 2018	$(1+1\%)\wedge(2/12)$ $-1 =$	0.17%	2	2	–	–
2. Hoarding	2,000,000	2,539,000	11,000	2,550,000	January 2019	$(1+1\%)\wedge(5/12)$ $-1 =$	0.42%	5	5	–	–
3. Site Formation Works	2,200,000	2,792,000	12,000	2,804,000	January 2019	$(1+1\%)\wedge(5/12)$ $-1 =$	0.42%	5	5	–	–
4. Foundation and Substructure	1,080,000	1,370,000	27,000	1,397,000	July 2019	$(1+1\%)\wedge(5/12)$ $\times(1+3\%)$ $(6/12)-1 =$	1.91%	11	5	6	–
5. Main Contract Works (including Superstructure, External Works & Landscaping)	36,830,000	46,747,000	1,403,000	48,150,000	June 2020	$(1+1\%)\wedge(5/12)$ $\times(1+3\%)\times$ $(1-1\%)\wedge(5/12)$ $-1 =$	3.00%	22	5	12	5
6. Preliminaries (15%)	6,500,000	Included	Included	Included							
7. Contingencies (10%)	5,000,000	Included	Included	Included							
Total	54,210,000	54,210,000	1,455,000	55,665,000			5.92%				

Example 4: The cot plan for completion of conceptual design for phase 2

Table 1.6 Example 4: The cost plan for completion of conceptual design for Phase 2 (storey height of G/F change from 3m to 5m)

(a) The C.F.A. Calculation for 1 House

Level	Area (CFA)
G/F	52 m2
1/F	52 m2
R/F	6 m2
Total	110 m2

Note:
Example 4 - Changing Storey height of G/F from 3m (phase 1) to 5m (phase 2) (where conceptual design is developed)

It is estimated by using phase 1 cost plan of conceptual design information with adjustment for change of the storey height combined with
(a) *HK$/m2 (buildup rate by elemental unit rate in Example 2 with adjustment where necessary) x C.F.A*
(b) *adjust elemental total by percentage of the height change*

(b) Elemental Cost Plan

Summary of Estimate (Preliminary Estimate)

Total Site Area: 3,620 m2
Total Gross Floor Area (GFA): 483 m2 (5,195 sq.ft.)
Total Construction Floor Area (CFA): 550 m2 = 110 m2 CFA × 5 houses
CFA/GFA ratio: 1.14

		Phase 1			Phase 2		
	CFA	Construction Cost	Unit Cost		Construction Cost	Unit Cost	
	(m2)	(HK$)	(HK$/m2 CFA)	(HK$/sq.ft. GFA)	(HK$)	(HK$/m2 CFA)	(HK$/sq.ft. GFA)
1. Site Investigation		600,000 *	1,091	115	600,000 *	1,091 *	115
2. Hoarding		2,000,000	3,636	385	2,000,000	3,636	385
3. Site Formation Works		2,200,000	4,000	423	3,300,000	6,000	635
4. Foundation and Substructure (pending engineer's input, assumed raft foundation for houses)		1,080,000	1,964	208	1,600,000	2,909	308
5. Superstructure	550	13,280,000	24,145	2,556	14,530,000	26,418	2,797
5.1 House A – 5 nos.	550	13,280,000	24,145	2,556	14,530,000	26,418	2,797

Table 1.6 Example 4: The cost plan for completion of conceptual design for phase 2 (storey height of G/F change from 3m to 5m) (Cont'd)

(b) Elemental cost plan

Summary of Estimate (Preliminary Estimate)

Total Site Area:	3,620 m2	
Total Gross Floor Area (GFA):	483 m2	(5,195 sq.ft)
Total Construction Floor Area (CFA):	550 m2	= 110 m2 CFA × 5 houses
CFA/GFA ratio:	1.14	

	CFA (m2)	Phase 1			Phase 2		
		Construction Cost (HK$)	Unit Cost (HK$/m2 CFA)	(HK$/sq.ft. GFA)	Construction Cost (HK$)	Unit Cost (HK$/m2 CFA)	(HK$/sq.ft. GFA)
6. External Works and Landscaping		23,550,000	42,818	4,533	23,550,000	42,818	4,533
6.1 External landscaping, paving and EVA		15,820,000	28,764	3,045	15,820,000	28,764	3,045
6.2 Paving/pedestrian walkway outside site boundary		1,100,000 *	2,000	212	1,100,000 *	2,000	212
6.3 Utilities within the site		5,700,000 *	10,364	1,097	5,700,000 *	10,364	1,097
6.4 Underground drainage		600,000 *	1,091	115	600,000 *	1,091	115
6.5 Utilities connections		330,000 *	600	64	330,000 *	600	64
Sub-total	550	42,710,000	77,655	8,221	45,580,000	82,873	8,774
7. Preliminaries (15% of Item 1–6)		6,500,000	11,818	1,251	6,900,000	12,545	1,328
8. Contingencies (10% of Item 1–7)		5,000,000	9,091	962	5,300,000	9,636	1,020
9. Fluctuation		1,455,000	2,645	280	1,519,000	2,762	292
TOTAL CONSTRUCTION COST (at January 2020 Price Level)		55,665,000	101,209	10,715	59,299,000	107,816	11,415

Remark:

* The costs are allowed figures for use in the presentation of the estimating exercise only and no further breakdown should be included.

Table 1.6 Example 4: The cost plan for completion of conceptual design for phase 2 (storey height of G/F change from 3m to 5m) (Cont'd)

Back-up Calculations

Descriptions	Quantity	Unit	Rate	Estimated Cost	Remarks
2. Hoarding			HK$	HK$	
a. Hoarding	265	m	7,500	1,987,500	remain unchanged
			Total	1,987,500	
3. Site Formation Works			Say	2,000,000	
a. Excavation; assumed 1.5m depth	5,430	m3	250	1,357,500	larger excavation volume due to thicker depth of raft footing
b. Cart away	5,430	m3	340	1,846,200	
			Total	3,203,700	
4. Foundation and Substructure			Say	3,300,000	
Total building footprint:	260 m2				
Total building footprint with raft foundation:	260 m2				
a. Raft foundation footing; assumed 1.5m thick	260	m2	6,000	1,560,000	As loading of super-structure increased, thicker raft footing is required
b. Blinding layer	260	m2	130	33,800	therefore rate $/m2 is increased
			Total	1,593,800	
5. Superstructure			Say	1,600,000	

Back-up calculations refer to next pages

6.1 External works and landscaping

Site Area: 3,620 m2

Less: G/F CFA of Houses: 260 m2

External landscaping, paving and EVA: 3,360 m2

a. External landscaping, paving and EVA including drainage, lighting, etc.	3,360	m2	4,000	13,440,000	
b. 2.5m high fence wall along site boundary; painted finish, design to budget	260	m	8,000	2,080,000	
c. Main entrance gate					
d. Signage		sum		100,000	remain unchanged
e. Guard house	1	sum		100,000	
		No.	100,000	100,000	
			Total	15,820,000	
			Say	15,820,000	

Table 1.6 Example 4: The cost plan for completion of conceptual design for phase 2 (storey height of G/F change from 3m to 5m) (Cont'd)

Elemental Breakdown for 5. Superstructure

Total Construction   550 m2 =   110 m2 × CFA ×
    Floor Area (CFA):   5houses

Elements	Phase 1		Phase 2		
	Elemental Total	Elemental Unit Cost	Cost for changing G/F height from 3m to 5m (Height of building changed from 8.425m to 10.425m high)		
			Elemental Total	Elemental Unit Cost	Remarks
			(HK$)	(HK$/m2)	Referring to Back-up Calculation of Superstructure
1. Carcase					
1.1 Frame and slab	2,770,000	5,036	3,230,000	* 5,873	* Refer to Note 1 in back-up calculations
					Cost of horizontal elements: 30%
					Cost of vertical elements: 70%
1.2 Internal walls	610,000	1,109	700,000	* 1,273	* Refer to Note 2 in back-up calculations
1.3 Doors and shutters	450,000	818	450,000	818	
Sub-total for Item 1	3,830,000	6,964	4,380,000	7,964	
2. Facade					* Refer to Note 3 in back-up calculations
2.1 External walls	1,230,000	2,236	1,510,000	* 2,745	
2.2 External wall finishes	520,000	945	640,000	* 1,164	* Refer to Note 4 in back-up calculations
2.3 Windows	690,000	1,255	690,000	1,255	
Sub-total for Item 2	2,440,000	4,436	2,840,000	5,164	
3. Finishings					
3.1 Roof finishes	70,000	127	70,000	127	
3.2 Floor finishes	810,000	1,473	810,000	1,473	
3.3 Internal wall finishes	590,000	1,073	730,000	* 1,327	* Refer to Note 5 in back-up calculations
3.4 Ceiling finishes	270,000	491	270,000	491	
3.5 Decor, graphics and signage	730,000	1,327	730,000	1,327	
Sub-total for Item 3	2,470,000	4,491	2,610,000	4,745	
4. Furniture and Fittings					
4.1 Built-in furniture	1,060,000	1,927	1,060,000	1,927	
4.2 Metal works and sundries	60,000	109	60,000	109	
4.3 Artwork	500,000	909	500,000	909	
4.4 Equipment	450,000	818	450,000	818	
4.5 Special light fittings	10,000	18	10,000	18	
Sub-total for Item 4	2,080,000	3,782	2,080,000	3,782	

Table 1.6 Example 4: The cost plan for completion of conceptual design for phase 2 (storey height of G/F change from 3m to 5m) (Cont'd)

Elemental Breakdown for 5. Superstructure

Total Construction Floor Area (CFA): 550 m2 = 110 m2 × CFA × 5houses

Elements	Phase 1		Phase 2		
	Elemental Total	Elemental Unit Cost	Cost for changing G/F height from 3m to 5m (Height of building changed from 8.425m to 10.425m high)		
			Elemental Total	Elemental Unit Cost	Remarks
			(HK$)	(HK$/m2)	Referring to Back-up Calculation of Superstructure
5. Building Services					
5.1 Sanitary fittings	200,000	364	200,000	364	
5.2 Plumbing and drainage	550,000	1,000	590,000*	1,073	
5.3 Electrical	830,000	1,509	880,000*	1,600	* Refer to Note 6 in back-up calculations
5.4 Fire services	280,000	509	300,000*	545	* Cost of vertical elements: ~ 28%
5.5 Mechanical ventilation and air conditioning	370,000	673	400,000*	727	
5.6 Gas	110,000	200	120,000*	218	
5.7 Builders work, profit and attendance (5%) of Item 5 Building Services	120,000	218	130,000	236	Adjusted accordingly
Sub-total for Item 5	2,460,000	4,473	2,620,000	4,764	
Total for Superstructure (Phase 2)	**13,280,000**	**24,145**	**14,530,000**	**26,418**	

Table 1.6 Example 4: The cost plan for completion of conceptual design for phase 2 (storey height of G/F change from 3m to 5m) (Cont'd)

Back-up Calculations for 5. Superstructure

Total Construction Floor Area (CFA):   550 m2   = 110 m2 × CFA × 5houses

Descriptions	Quantity	Unit	Rate	Phase 1 Estimated Cost	Phase 2 Estimated Cost	Phase 2 Remarks — Cost for changing G/F height from 3m to 5m (Height of building changed from 8.425m to 10.425m high)
			(HK$)	(HK$)	(HK$)	
1.1 Frame and slabs						
Horizontal elements						
a. Reinforced concrete, grade 45D; to slab and beam	133	m3	1,300.00	172,900		
b. Rebar to slab and beam (230kg/m3)	30,590	kg	12.00	367,080		
c. Formwork to slab and beam	665	m2	350.00	232,750		
d. Allow 5% for miscellaneous				40,000		
		Total for Horizontal Elements:		812,730		Note 1
						Calculated cost of:-
						1. horizontal elements:
						812,946 / 2,770,000 = ~30%
Vertical elements						
a. Reinforced concrete, grade 45D; to structural wall	227	m3	1,300.00	295,100		2. vertical elements:
b. Rebar to structural wall (280kg/m3)	63,560	kg	12.00	762,720		1,951,056 /
c. Formwork to structural wall	2,268	m2	350.00	793,800		2,770,000 = ~70%
d. Allow 5% for miscellaneous				100,000		
		Total for Vertical Elements:		1,951,620		Adjusted percentage:
						30% + 70% / 8.425 x 10.425 = 117%
			Total	2,764,350	3,234,290 *	Adjusted amount:
			say	2,770,000	3,230,000	2,764,350 x 117%

Table 1.6 Example 4: The cost plan for completion of conceptual design for phase 2 (storey height of G/F change from 3m to 5m) (Cont'd)

Back-up Calculations for 5. Superstructure

Total Construction Floor Area (CFA):   550 m2   = 110 m2 × CFA × 5houses

Descriptions	Quantity	Unit	Rate	Phase 1 Estimated Cost	Phase 2 Cost for changing G/F height from 3m to 5m (Height of building changed from 8.425m to 10.425m high)	
					Estimated Cost	Remarks
			(HK$)	(HK$)	(HK$)	
1.2 Internal walls						
a. 75mm thick concrete block wall	546	m2	700.00	382,200	473,928.00 *	Note 2
b. Glass partition for shower	45	m2	5,000.00	225,000	225,000	Adjusted percentage:
			Total	607,200	698,928	100% / 8.425 x 10.425 = 124%
			Say	**610,000**	**700,000**	Adjusted amount: 382,200 x 124%
1.3 Doors and shutters						
a. Double leaf glass doors to flat entrances	5	no.	20,000.00	100,000	100,000	
b. Single leaf to bedroom	20	no.	6,000.00	120,000	120,000	
c. Single leaf doors to bathroom	10	no.	6,000.00	60,000	60,000	remain unchanged
d. Single leaf doors to kitchen	10	no.	7,000	70,000	70,000	
e. Single leaf doors to living/dining room	10	no.	6,000	60,000	60,000	
f. Single leaf doors to roof	5	no.	7,000.00	35,000	35,000	
			Total	445,000	445,000	
			Say	**450,000**	**450,000**	
2.1 External walls						
a. 200mm thick reinforced concrete wall	1,079	m2	1,100.00	1,186,900	1,471,756 *	Note 3
b. 150mm thick reinforced concrete parapet wall	33	m2	1,100.00	36,300	36,300	Adjusted percentage:
			Total	1,223,200	1,508,056	100% / 8.425 x 10.425 = 124%
			Say	**1,230,000**	**1,510,000**	Adjusted amount: 1,186,900 x 124%

(Continued)

Table 1.6 Example 4: The cost plan for completion of conceptual design for phase 2 (storey height of G/F change from 3m to 5m) (Cont'd)

Back-up Calculations for 5. Superstructure

Total Construction Floor Area (CFA): 550 m2 = 110 m2 × CFA × 5houses

Descriptions	Quantity	Unit	Rate	Phase 1 Estimated Cost	Phase 2 Cost for changing G/F height from 3m to 5m (Height of building changed from 8.425m to 10.425m high) Estimated Cost	Remarks
			(HK$)	(HK$)	(HK$)	
2.2 External wall finishes						
a. Ceramic tiles (PC. HK$50/m2) to external wall	1,079	m2	450.00	485,550	602,082.00 *	Note 4
b. Ceramic tiles (PC. HK$50/m2) to parapet wall	66	m2	450.00	29,700	29,700	Adjusted percentage:
			Total	515,250	631,782	100% / 8.425 x 10.425 = 124%
			Say	520,000	640,000	Adjusted amount: 485,550 x 124%
2.3 Windows						
a. Window to bedroom	56	m2	3,500.00	196,000	196,000	
b. Window to bathroom	6	m2	3,500.00	21,000	21,000	
c. Window to kitchen	11	m2	3,500.00	38,500	38,500	
d. Window to living/dining room	28	m2	3,500.00	98,000	98,000	remain unchanged
e. Window to internal staircase	56	m2	6,000.00	336,000	336,000	
			Total	689,500	689,500	
			Say	690,000	690,000	
3.1 Roof finishes						
a. Homogeneous tiles including cement sand screed, insulation and waterproofing	46	m2	1,300.00	59,800	59,800	
b. Waterproofing and insulation only	6	m2	500.00	3,000	3,000	remain unchanged
c. Allow for skirting (10%)		sum		6,280	6,280	
			Total	69,080	69,080	
			Say	70,000	70,000	

Table 1.6 Example 4: The cost plan for completion of conceptual design for phase 2 (storey height of G/F change from 3m to 5m) (Cont'd)

Back-up Calculations for 5. Superstructure
Total Construction Floor Area (CFA): 550 m2   = 110 m2 × CFA × 5houses

Descriptions	Quantity	Unit	Rate	Phase 1 Estimated Cost	Phase 2 Estimated Cost	Remarks
			(HK$)	(HK$)	(HK$)	Cost for changing G/F height from 3m to 5m (Height of building changed from 8.425m to 10.425m high)
3.2 Floor finishes						
a. Natural stone (P.C. HK$1,500/m2) to house entrance	33	m2	1,400.00	46,200	46,200	
b. Stone finishes (P.C. HK$600/m2) to bedroom	108	m2	1,400.00	151,200	151,200	
c. Stone finishes (P.C. HK$600/m2) to bathroom	29	m2	1,400.00	40,600	40,600	
d. Stone finishes (P.C. HK$600/m2) to kitchen	38	m2	1,400.00	53,200	53,200	
e. Stone finishes (P.C. HK$600/m2) to living/dining room	205	m2	1,400.00	287,000	287,000	remain unchanged
f. Stone finishes (P.C. HK$600/m2) to internal staircase; including nosing tiles	107	m2	1,400.00	149,800	149,800	
g. Allow for skirting (10%)		sum		72,800	72,800	
			Total	800,800	800,800	
			Say	810,000	810,000	
3.3 Internal wall finishes						*Note 5 Adjusted percentage:
a. Plaster with emulsion paint to house entrance	147	m2	200.00	29,400		100% / 8.425 x 10.425 = 124%
b. Plaster with emulsion paint to bedroom	592	m2	200.00	118,400		
c. Glazed ceramic tiles (P.C. HK$200/m2) to bathroom	207	m2	650.00	134,550		Adjusted amount:
d. Glazed ceramic tiles (P.C. HK$ 150/m2) to kitchen	240	m2	500.00	120,000		587,950 x 124%
e. Plaster with emulsion paint to living/dining room	662	m2	200.00	132,400		

(Continued)

Table 1.6 Example 4: The cost plan for completion of conceptual design for phase 2 (storey height of G/F change from 3m to 5m) (Cont'd)

Back-up Calculations for 5. Superstructure
Total Construction Floor Area (CFA):　　　550 m2　　　= 110 m2 × CFA × 5houses

Descriptions	Quantity	Unit	Rate		Phase 1 Estimated Cost		Phase 2 Cost for changing G/F height from 3m to 5m (Height of building changed from 8.425m to 10.425m high)	
							Estimated Cost	Remarks
				(HK$)	(HK$)		(HK$)	
f. Plaster with emulsion paint to internal staircase	266	m2	200.00		53,200		729,058 *	
			Total		587,950			
			Say		**590,000**		**730,000**	
3.4 Ceiling finishes								
a. Plaster with emulsion paint to house entrance	33	m2	200.00		6,600		6,600	
b. Plaster with emulsion paint to bedroom	108	m2	200.00		21,600		21,600	
c. Gypsum board suspended ceiling to bathroom	29	m2	1,500.00		43,500		43,500	
d. Gypsum board suspended ceiling to kitchen	38	m2	1,500.00		57,000		57,000	
e. Plaster with emulsion paint to living/dining room	205	m2	200.00		41,000		41,000	remain unchanged
f. Plaster with emulsion paint to internal staircase	107	m2	200.00		21,400		21,400	
g. Allow for bulkhead	5	house	15,000.00		75,000		75,000	
			Total		266,100		266,100	
			Say		**270,000**		**270,000**	
3.5 Decor, graphics and signage								
a. Decor, graphics and signage		sum			730,000		730,000	remain unchanged
			Total		730,000		730,000	
			Say		**730,000**		**730,000**	
4.1 Built-in furniture								
a. Kitchen cabinet with worktop	10	no	90,000.00		900,000		900,000	
b. Vanity counter with marble countertop and mirror cabinet to bathroom	10	no	15,000.00		150,000		150,000	remain unchanged
c. Letter box	5	house	1,500.00		7,500		7,500	
			Total		1,057,500		1,057,500	
			Say		**1,060,000**		**1,060,000**	

Table 1.6 Example 4: The cost plan for completion of conceptual design for phase 2 (storey height of G/F change from 3m to 5m) (Cont'd)

Back-up Calculations for 5. Superstructure
Total Construction Floor Area (CFA):   550 m2   = 110 m2 × CFA × 5houses

Descriptions	Quantity	Unit	Rate	Phase 1 Estimated Cost	Phase 2 — Cost for changing G/F height from 3m to 5m (Height of building changed from 8.425m to 10.425m high) Estimated Cost	Remarks
			(HK$)	(HK$)	(HK$)	
4.2 Metal works and sundries						
a. Metal works and sundries	550	m2	100.00	55,000	55,000	remain unchanged
			Total	55,000	55,000	
			Say	60,000	60,000	
4.3 Artwork						
a. Artwork at house entrance	5	no	100,000.00	500,000	500,000	remain unchanged
			Total	500,000	500,000	
			Say	500,000	500,000	
4.4 Equipment						
a. Kitchen appliances	10	house	40,000.00	400,000	400,000	remain unchanged
b. Electric water heater to kitchen	10	house	5,000.00	50,000	50,000	
			Total	450,000	450,000	
			Say	450,000	450,000	
4.5 Special light fittings						
a. Special lighting fittings for house entrance	33	m2	250.00	10,000	10,000	remain unchanged
			Total	10,000	10,000	
			Say	10,000	10,000	
5.1 Sanitary fittings						
a. Sanitary fittings to bathroom	10	no	15,000.00	150,000	150,000	remain unchanged
b. Allow for bathroom accessories	10	set	5,000.00	50,000	50,000	
			Total	200,000	200,000	
			Say	200,000	200,000	

(Continued)

Table 1.6 Example 4: The cost plan for completion of conceptual design for phase 2 (storey height of G/F change from 3m to 5m) (Cont'd)

Back-up Calculations for 5. Superstructure

Total Construction Floor Area (CFA):   550 m2   = 110 m2 × CFA × 5houses

Descriptions	Quantity	Unit	Rate	Phase 1 Estimated Cost	Phase 2 Estimated Cost	Phase 2 Remarks — Cost for changing G/F height from 3m to 5m (Height of building changed from 8.425m to 10.425m high)
			(HK$)	(HK$)	(HK$)	
5.2 Plumbing and drainage						
a. Plumbing and drainage	550	m2	1,000.00	550,000	586,558 *	* Note 6
			Total	550,000	586,558	Adjusted of vertical elements around 28%:
			Say	550,000	590,000	Int. wall finishes/Total finishings 730,000/2,610,000 = 28% Adjusted amount: 550,000 x 72% + 28%/8.425 x 10.425
5.3 Electrical						
a. Electrical	550	m2	1,500.00	825,000	879,837 *	Adjusted amount:
			Total	825,000	879,837	825,000 x 72% + 28%/8.425 x 10.425
			Say	830,000	880,000	
5.4 Fire Services						
a. Fire services	550	m2	500.00	275,000	293,279 *	Adjusted amount:
			Total	275,000	293,279	275,000 x 72% + 28%/8.425 x 10.425
			Say	280,000	300,000	
5.5 Mechanical ventilation and air conditioning						
a. Split type air conditioning to house entrance	5	no	20,000.00	100,000		
b. Thermal ventilator to bedroom	20	no	6,000.00	120,000		
c. Ventilation fans to bathroom and kitchen	20	no	6,000.00	120,000		
d. Allow for bulkhead of mechanical ventilation	5	house	6,000.00	30,000		
			Total	370,000	394,593 *	Adjusted amount:
			Say	370,000	400,000	370,000 x 72% + 28%/8.425 x 10.425

Table 1.6 Example 4: The cost plan for completion of conceptual design for phase 2 (storey height of G/F change from 3m to 5m) (Cont'd)

Back-up Calculations for 5. Superstructure
Total Construction Floor Area (CFA): 550 m2 = 110 m2 × CFA × 5houses

Descriptions	Quantity	Unit	Rate	Phase 1 Estimated Cost	Phase 2 Estimated Cost	Remarks
					Cost for changing G/F height from 3m to 5m (Height of building changed from 8.425m to 10.425m high)	
			(HK$)	(HK$)	(HK$)	
5.6 Gas						
a. Gas	550	m2	200.00	110,000	117,312 *	Adjusted amount: 111,000 x 72% + 28%/8.425 x 10.425
			Total	110,000	117,312	
			Say	110,000	120,000	
5.7 Builders work, profit and attendance						
(5%) of Item 5 Building Services		sum		117,000	124,500 *	Adjusted accordingly
			Total	117,000	124,500	
			Say	120,000	130,000	
Total of Superstructure				13,280,000	14,530,000	

Table 1.6 Example 4: The cost plan for completion of conceptual design for phase 2 (storey height of G/F change from 3m to 5m) (Cont'd)

FORECAST OF CONSTRUCTION COST FLUCTUATION
Price Level in Estimate: July 2018

* Assumed fluctuation percentage as follows:
January 2018 to December 2018: +1%
January 2019 to December 2019: +3%
January 2020 to December 2020: -1%

Phase 2

Works Packages	Anticipated Contract Sum	Anticipated Contract Sum (Including Preliminaries and Contingencies)	Anticipated Cost of Inflation up to Tender Date	Total Construction Cost Including Inflation	Anticipated Tender Award Date	Percentage Calculation Formula	Fluctuation Percentage	Fluctuation Period	Year 2018	Year 2019	Year 2020
	(HK$)	(HK$)	(HK$)	(HK$)			(% p.a.)	(Month)	+1%	+3%	-1%
1. Site Investigation	600,000	761,000	2,000	763,000	October 2018	$(1+1\%)^{(2/12)} - 1 =$	0.17%	2	2	-	-
2. Hoarding	2,000,000	2,535,000	11,000	2,546,000	January 2019	$(1+1\%)^{(5/12)} - 1 =$	0.42%	5	5	-	-
3. Site Formation Works	3,300,000	4,183,000	18,000	4,201,000	January 2019	$(1+1\%)^{(5/12)} - 1 =$	0.42%	5	5	-	-
4. Foundation and Substructure	1,600,000	2,028,000	39,000	2,067,000	July 2019	$(1+1\%)^{(5/12)} \times (1+3\%)^{(6/12)} - 1 =$	1.91%	11	5	6	-
5. Main Contract Works (including Superstructure, External Works & Landscaping)	38,080,000	48,273,000	1,449,000	49,722,000	June 2020	$(1+1\%)^{(5/12)} \times (1+3\%) \times (1-1\%)^{(5/12)} - 1 =$	3.00%	22	5	12	5

(Continued)

Table 1.6 Example 4: The cost plan for completion of conceptual design for phase 2 (storey height of G/F change from 3m to 5m) (Cont'd)

FORECAST OF CONSTRUCTION COST FLUCTUATION
Price Level in Estimate: July 2018

* Assumed fluctuation percentage as follows:
January 2018 to December 2018: +1%
January 2019 to December 2019: +3%
January 2020 to December 2020: -1%

Works Packages	Anticipated Contract Sum	Anticipated Contract Sum (Including Preliminaries and Contingencies)	Anticipated Cost of Inflation up to Tender Date	Total Construction Cost Including Inflation	Anticipated Tender Award Date	Percentage Calculation	Fluctuation Percentage (% p.a.)	Fluctuation Period (Month)	Year 2018 +1%	Year 2019 +3%	Year 2020 −1%
	Phase 2										
	(HK$)	(HK$)	(HK$)	(HK$)							
6. Preliminaries (15%)	6,900,000	Included	Included	Included							
7. Contingencies (10%)	5,300,000	Included	Included	Included							
Total	57,780,000	57,780,000	1,519,000	59,299,000		Formula	5.92%				

2   Tendering and tender documentation

Learning outcomes

Upon completion of this chapter, you should be able to do the following.

1   Understand the principle of tendering in construction.
2   Understand the typical tendering procedure for a construction project.
3   Know how to prepare a tender.
4   Understand the composition of a set of tender documents and their functions.

2.1 Introduction

Competitive tendering is used in the construction industry to select a capable contractor for a construction project. In Hong Kong, the tendering stage is usually short, and the quantity surveyor must prepare many deliverables within a brief period of time so that the construction contracts can be awarded and the construction projects started in a timely manner. Tender documents are an essential part of a successful tender, as they become the documents used for post-contract contractual and financial management. If the information provided in the tender documents is sufficiently detailed and substantive, the contractor can be certain about the provisions in his offer and can meet the specified requirements. If the information is not sufficient, additional allowances are typically needed to address doubts and ambiguities; this will be reflected in the tender price and will increase the difficulties in post-contract cost monitoring and control. This chapter reviews the tendering process and discusses the role and duties of the quantity surveyor in tender preparation. Some case studies are then provided.

2.2 Tendering in construction

Competitive tendering is common in the construction industry. The employer usually uses this method to select a capable contractor for a construction project. Figure 2.1 illustrates the typical tendering procedure for a construction project under a traditional tendering procurement strategy. The overall procedure is comprised of four key stages: 1) tender invitation; 2) tender return; 3) tender examination; and 4) tender award. This chapter focuses on the documents that the quantity surveyor prepares when inviting a competitive tender. Chapter 3 covers the remaining stages from tender return to tender award.

2.2.1 Tender invitation

A tender invitation is not an offer from the employer, but an invitation to submit a tender price as a bid. The invitation letter should indicate a willingness to receive tenderers' offers

DOI: 10.1201/9781003212355-2

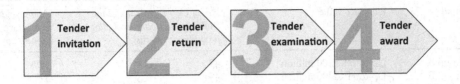

Figure 2.1 Stages in a typical tendering procedure

and that the employer is free to accept or reject any tender received. Open tendering and selective tendering are both common in Hong Kong, but their use differs in the public and private sectors. Some public projects in Hong Kong use open tendering, which are normally made via gazette notices and/or posted tender notices on the Internet. The gazette is an official publication of the Hong Kong Government, which is normally published on Fridays. It consists of seven sections; public tenders are included in the main gazette and include details such as the scope of the work, project commencement date(s) and contract period. Tender return information is also included for tenderers' reference. Selective tendering is also common in the public sector. For example, the Architectural Services Department (ArchSD) uses selective tendering such that only qualified contractors on the List of Approved Contractors for Public Works are invited to tender. This list is comprised of contractors who are approved to carry out public works in one or more of the five major categories of building and civil engineering works. These qualified contractors are required to meet the financial, technical and management criteria for admission and retention on the approved lists. Table 2.1 shows the three groups of approved contractors for public works and their corresponding contract values.

2.2.2 Pre-qualification

Pre-qualification is a commonly used form of selective tendering in the private sector. Contractors who express interest in tendering for the proposed project are invited to participate in a pre-qualification exercise that creates a shortlist of contractors who are technically and financially capable of undertaking the proposed construction project. These contractors will then be invited to submit tender bids. During the pre-qualification process, each contractor's technical and financial characteristics are evaluated, including their technical experience, past performance record, labour and plant resources, current workload, management structure, safety and quality management, and company financial standing and record. Finally, a tenderer list is compiled. The project team should keep the tenderer list confidential.

2.2.3 Tender documents

To prepare an offer, which if accepted will form the basis of a contract between the tenderer and the employer, the tenderer must be provided with sufficient information about the project and the requirements for executing the tendered works. Tender documents should include all of the information that tenderers need to prepare their bids. The completeness of the information determines the accuracy of the tenders.

A tender typically contains two parts: the front part and the pricing part. The front part generally includes information about the tendering exercise, contract conditions and project

Table 2.1 List of approved contractors for public works (as of July 2020)

Group	Contract Value	No. of Contractors under the Buildings Category (including probationary status)	Approximate Percentage of Contractors
Group A	Up to HK$100 million	49	31%
Group B	Up to HK$300 million	49	31%
Group C	Exceeding HK$300 million	61	38%

specifications. The details of the individual documents are discussed in the next sections of this chapter. It is essential for tenderers to understand the rules of tendering as well as the technical and contractual requirements of the construction project.

The pricing part of the tender includes a schedule or schedules that may include the quantity of works to be completed in recognised units of measurement (which varies with the type of construction contract used) and other associated items such as preliminaries, preambles, prime cost sums and provisional sums. Tenderers can price the schedule(s) to become their estimated construction cost (i.e., the tender bid).

It is common practice to issue tenders in loose-leaf booklets consisting of documents with hole-punched pages. This allows tenderers to separate the tender documents by work trade or package to obtain competitive quotations from sub-contractors. Alternatively, a soft copy of the tender can be given to tenderers using an electronic tender dissemination method. Tender drawings prepared by design consultants are also distributed to tenderers. The tendering period is determined based on the project scale and the overall tender programme. However, it should be sufficient to ensure that tenderers have time to obtain competitive quotations for materials and sublet work.

2.3 Front part

The front part of the tender generally consists of four sections: Conditions of Tender, Form of Tender, Conditions of Contract and Specification. The Conditions of Tender and Form of Tender sections deal with tendering issues, and the Conditions of Contract and Specification outline the contractual requirements and project-based information for the project. The front part provides tenderers with the conditions of the tendering exercise and the project requirements, such as contract conditions, specified performance quality, site facilities and general obligations.

2.3.1 Conditions of tender

Tenderers must comply with the conditions stated in the Conditions of Tender. When drafting the Conditions of Tender, the quantity surveyor must communicate with the employer and the project team, especially the architect, on key items such as the location of the tender return and the requirements for submitting supplementary information for tender evaluation. This section includes three types of tender information that critically influence tenderers' pricing: the tendering requirements, the employer's right to tender and the tender analysis. Some typical conditions under these three categories are listed in Table 2.2. The information in the Conditions of Tender is important to tenderers, who must read them carefully to determine accurate tender pricing.

Table 2.2 Typical conditions stated in the conditions of tender and their purposes

Types and Purposes of Conditions of Tender	Example Conditions
A. Tendering Requirements	1. Tenderers to check the tender documents, including tender drawings issued 2. List of supplementary information to be submitted by tenderers with their tenders 3. Time and location for tender submission 4. Period during which tenderers can raise queries regarding the tender 5. Proper tender pricing, including pricing consistency and no unauthorized alteration, currency, etc. 6. No disclosure of private and confidential project information
B. Employer's Rights in Tendering	1. The employer is not bound to accept either the lowest or any tender submitted 2. Late or non-compliant tenders will not be considered 3. No monetary compensation will be provided to tenderers regarding their tender preparation
C. Tender Analysis	1. Methods for rectifying arithmetical errors found in pricing documents such as bills of quantities 2. Issue of tender clarification/notification to tenderers after tender submission

2.3.2 Form of tender

In construction tendering, a tender offer is made by the tenderer to the employer. The offer should be sufficiently specific in terms of tender price and obligations and should be made with the expectation that it will be legally binding upon acceptance. Thus, the Form of Tender allows a tenderer to respond to an employer's tender invitation and, if the tender offer is accepted, to execute the construction works in accordance with the terms and conditions of the contract within the contract period. The Form of Tender is a pro forma document that contains the following essential information:

1   company details (e.g., company name and registered address);
2   business registration certificate number and its expiry date;
3   tender sum (to be written in numbers and words);
4   signature of person authorised to sign the tender and his/her position in the company; and
5   signature of witness and his/her personal details.

Tender sums should remain open for acceptance for a specified period; acceptance is not possible after the expiry date. Thus, the Form of Tender should include a tender validity period. If the employer does not accept one of the tenders during this period, the quantity surveyor, after seeking the employer's instruction, should seek written consent from each of the tenderers to extend the tender validity period. Tenderers are also required to confirm that no unauthorised alterations have made to the tender documents and are required to obtain a surety bond if the tender offer is accepted.

The employer's requirements for the tender award are also stated in the Form of Tender section. The following statements are typical examples about the conditions of tender acceptances and binding contracts.

1   The Employer reserves the right and is not bound to accept the lowest or any tender submitted.
2   Pending the signing of the formal agreement, the tender together with written acceptance from the Employer will constitute a binding agreement once the tender offer has been accepted.

Typically, all tenderers are required to submit tender prices on an equal basis. However, if alternatives are invited, tenderers will be allowed to submit alternative tender prices. For example, tenderers may be asked to propose an alternative contract period. In this case, two Forms of Tender will be provided in the tender documents, i.e., Forms of Tender A and B. Form of Tender A is a base tender (i.e., the tender for the specified contract period), and Form of Tender B is an alternative tender (i.e., the tender for a proposed alternative contract period). In this case, each tenderer will submit two tender prices, and the employer can award the contract based on any of them.

2.3.3 General and special conditions of contract

In the private sector of Hong Kong's building industry, most contracts use the general conditions given in the 2005 edition of the Standard Form of Building Contract published by the Hong Kong Institute of Architects, the Hong Kong Institute of Construction Managers and the Hong Kong Institute of Surveyors (SFBC). In the public sector, the most common conditions are those in the Hong Kong Government General Conditions of Contract 1999 edition (HKGGCC), published by the Hong Kong government. The SFBC and HKGGCC are both open standard documents. Tenderers are free to consult them, so it is not common practice to include the SFBC or HKGGCC in tender documents. However, the SFBC and HKGGCC are usually modified and supplemented in a tender in a section entitled Special or Supplementary Conditions of Contract (SCC), which allows the quantity surveyor to modify the conditions to meet the needs of a particular employer or a set of project requirements. An SCC should be included in the tender, and it may substantially affect the tender price. For example, in a private sector project, the quantity surveyor usually drafts the SCC based on a standard SCC maintained by the quantity surveyor's or the employer's company. The quantity surveyor should clearly inform the tenderers of the amendments to the SFBC, including which contract clauses have been deleted/amended and what new contract provisions have been stipulated. In addition, the quantity surveyor should communicate with the employer to ensure that the SCC sufficiently addresses the employer's concerns and project particulars. Some private developers prefer to use their own construction contracts instead of the SFBC, in which case it is necessary to incorporate the full set of contract conditions (including both general conditions and an SCC) into the tender as not all contractors will be familiar with the non-standard contract conditions.

2.3.4 Specification

Broadly speaking, a specification has two parts: preliminaries and technical specifications. A preliminaries specification provides an overview of important financial matters related to the project and is not confined to any particular trade. In general, the preliminaries specification also outlines requirements that affect tenderers' pricing. Tenderers must price the preliminaries specification into a preliminaries bill. The preliminaries bill lists the items required for each clause in the preliminaries specification. In the Hong Kong Standard Method of Measurement of Building Works Fourth Edition (HKSMM4), the preliminaries specification document has three core sections: (i) preliminary particulars; (ii) conditions of contract; and (iii) general matters. The preliminary particulars section covers site access and scope of work. The conditions of contract section covers the form of the construction contract and the particulars to be included in its appendix. The general matters section outlines the requirements related to the contractor's general obligations, temporary work, safety, insurance and sub-contracting. It is common practice for the quantity surveyor to issue a list of queries to the architect or project team asking them to provide confirmation of contract particulars that will form the basis of the preliminaries specification.

A technical specification is a document prepared by design consultants that specifies the required quality of the materials and workmanship. It also includes any constraints on construction methods. Most consultant firms include a set of general specifications in the tender that outline the general requirements for the project and then supplement this document with the requirements for the particular project.

2.4 Pricing part

In the pricing part of the tender, tenderers calculate a bid by pricing the unit rates in the bills of quantities or the schedule of quantities and rates. A set of preambles is included to provide essential information about item and rate coverage. Prime cost and provisional sums are also included.

2.4.1 Bills of quantities

The bills of quantities (BQ) is a schedule that sets out the quantities of work in recognised units of measurement as outlined in the Standard Method of Measurement. The BQ is an important part of a tender as it enables every tenderer to calculate pricing on an equal basis. The BQ sets out the descriptions and quantities of construction tasks based on the drawings and the requirements laid down in the specifications, and thus reduces the risk in tender pricing. The BQ includes a comprehensive, itemised list of the contract works, which can provide a good basis for evaluating a tender and future variations and cost analysis. The BQ consists of various bills for tenderers' pricing, including preliminaries, preambles, measured works, daywork, prime cost, and provisional sums. The tender sum is the total cost of executing all of the items listed in these bills. BQ measurement is commonly used for lump sum contracts.

2.4.2 Bill of preliminaries

The bills of preliminaries should be read in conjunction with the preliminaries specification. Tenderers should price the items that are deemed to apply to the whole project, and no adjustment will be made for any additions or omissions to the contract works by the architect. The cost breakdown of the preliminaries will form the basis for post-contract cost management, such as valuation of interim payments and assessments of site overhead under claims of disruption or prolongation. The quantity surveyor should therefore produce a bill of preliminaries that is sufficiently detailed to achieve the aforementioned purposes. If a tenderer prices a lump sum, the quantity surveyor should ask them to provide an itemised breakdown for tender consideration.

2.4.3 Bill of preambles

Any departures from the HKSMM4 in the measurement should be explicitly stated in the preambles. The bill of preambles includes any qualifications and clarifications to the HKSMM4, and its incorporation into the BQ is required under the HKSMM4. The preambles are part of the BQ descriptions. Because of length limitations, the full requirements of the HKSMM4 are not repeated in the BQ; thus, the required information is given in the preambles. The preambles also include the item coverage for pricing the unit rates. As recommended by the HKSMM4, the following costs associated with the work are included in the measured items unless otherwise stated and are thus included in the unit rates priced by the tenderers:

1   labour and all costs in connection therewith;
2   materials, goods and all costs in connection therewith (e.g., conveyance, delivery, unloading, storing, returning packages, handling, hoisting, lowering);
3   fitting and fixing materials and goods in position;
4   use of plant;
5   all cutting and waste; and
6   establishment charges, overhead charges and profit.

In practice, the bill of preambles is a standard document in tenders that should be amended and updated as necessary. It is formatted according to the HKSMM4 guidelines. The quantity surveyor should check the preambles before measuring the BQ quantities, because the standard measurement rules may be changed by amendments and qualifications included in a project's preambles. In some situations, such amendments or qualifications would be included as special preambles. Accordingly, the bill of preambles is a crucial component of tenderers' pricing, because the cost of its overall effect must be included in the unit rates of the measured items in the BQ.

2.4.4 Bills of measured works

The BQ provides the descriptions and quantities of the measured works as outlined in the HKSMM4. In a lump sum contract, with the exception of items measured as provisional, all of the BQ quantities are measured as firm quantities and form part of the construction contract. No adjustment should be made unless ordered by the architect. According to the 2005 edition of the Agreement and Schedule of Conditions of Building Contract published by the Hong Kong Institute of Architects, the Hong Kong Institute of Construction Managers and the Hong Kong Institute of Surveyors, any error found in the BQ should be rectified and corrected by the quantity surveyor and deemed to be a variation. Provisional quantities should be allowed in the BQ in case design details for those works were insufficiently detailed or unavailable during the BQ preparation. Generally, the quantity surveyor provisionally allows approximate quantities subject to final remeasurement. In the remeasurement process, provisional quantities are adjusted based on finalised drawings provided by design consultants during the construction or final account stage. The BQ forms the major part of the tender pricing, and the unit rates contained in the BQ are used for the valuation of variations. Any deviations from the measurement rules of the HKSMM4 in the BQ preparations should be stated in the preambles.

The bills in the BQ cover the contract works of the tendered project, and the quantities are measured according to the HKSMM4. This fourth edition was updated in 2018 and published by the Hong Kong Institute of Surveyors. It is currently in use in Hong Kong. The HKSMM4 includes the measurement rules for building works and building services installations. The HKSMM4 classifies the various types of works and provides a predetermined order for the works presented in the BQ. Two BQ formats are commonly used by quantity surveyors.

1   Elemental bill format
    In the elemental bill format, BQ items are grouped according to their location in the building. Every element, such as a floor, external wall or roof, is an integral part of the building and performs a certain design function. The BQ items for each element are measured and presented in a work order as stipulated in the HKSMM4.
2   Work section (trade) bill format
    In the work section bill format, items are grouped by work section as stipulated in the HKSMM4. A work section bill can also be subdivided by its elements; this kind of bill is called a sectionalised work section bill.

2.4.5 Bill of provisional and prime cost sums

A separate bill covers the provisional and prime costs of the tendered project. Provisional sums are allowed for work carried out by the contractor but do not provide enough detail for the measurement needed for the BQ preparation. A provisional sum is also allowed as a contingency to cover the future costs of minor design changes and improvements as well as any unanticipated risks and uncertainties that may arise. The quantity surveyor should estimate the provisional sums based on the latest cost plan and information provided by the design consultants. The architect should issue instructions for the expenditure of provisional sums, which may be expended in whole or in part. Unexpended provisional sums should be deducted from the contract sum in the final account.

Prime cost (PC) sums cover specialist work carried out by the nominated sub-contractors or specific materials supplied by the nominated suppliers. As with provisional sums, the quantity surveyor should estimate the amounts of PC sums in consultation with the building services engineer, who may be responsible for the budget allowance of the nominated sub-contracts in charge of building services installations. In addition, the quantity surveyor should include separate items in the bill of provisional and PC sums so that the contractor can price the profit and attendance for each PC sum. During the construction process, the contractor is required to provide the nominated sub-contractors and nominated suppliers with all of the elements of site attendance, including welfare facilities, safety provisions and temporary power and water supplies. The contractor is thus required to price these site facilities as attendance costs against each PC sum stated in the bills of provisional and PC sums. Figure 2.2 shows an example of a bill of provisional and PC sums.

Item	Description	HK$
	BILL NO. 12 – Provisional and Prime Cost Sums	
	PROVISIONAL SUMS The following Provisional Sums may be expended in whole or in part as directed by the Architect or wholly deducted from the Contract Sum if not required	
A	Provide the sum of HK$1,200,000 for signage	1,200,000.00
B	Provide the sum of HK$1,500,000 for architectural features	1,500,000.00
C	Provide the sum of HK$1,000,000 for the builder's work in connection with building services	1,000.000.00
D	Provide the sum of HK$1,800,000 for contingencies	1,800,000.00
	PRIME COST SUMS Provide the following Prime Cost Sums for the Supply and Fixing of Works by Specialists who shall be nominated by the Architect and who are declared to be Nominated Sub-contractors to the Contractor in accordance with the Conditions of Contract	
E	Provide the sum of HK$25,000,000 for electrical installation	25,000,000.00
F	Add ___ % for profit	
G	Add for attendance, etc. as described	

Figure 2.2 Example of a bill of provisional and PC sums

2.4.6 General summary

The General Summary of the BQ summarises the total of each bill. It has two key purposes. First, it allows tenderers to summarise the bill's components and present a total (i.e., a tender bid), which should be the same as the bid in the Form of Tender. Second, the authorised person from the tenderer's company will sign the General Summary page and stamp it with a company chop. The contractor's senior management then confirms the final offer prior to signing the Form of Tender and submitting the tender bid. Figure 2.3 shows an example of a General Summary of the BQ.

2.4.7 Schedule of quantities and rates

A schedule of quantities and rates (SQR) is another type of pricing document. Tenderers are required to provide a lump sum for completing the construction work that is outlined in the drawings and specifications. Traditionally, a schedule of rates is used to list the cost of various work items without quantities. Tenderers should read and study the drawings and specifications carefully to acquaint themselves with the full scope of the contract. A schedule of rates is not meant to present the complete scope of the work in a tendered project. Tenderers should enter their own quantities into the schedule of rates and any additional items that they consider necessary based on the tender drawings and specifications. These quantities are

THE SUPERSTRUCTURE MAIN CONTRACT FOR
THE PROPOSED RESIDENTIAL DEVELOPMENT
GENERAL SUMMARY

Bill No.		*Page No.*	*HK$*
1.	Preliminaries	B1/SUM	
2.	Preambles	–	
3.	Excavation	B3/SUM	
4.	Concrete Works	B4/SUM	
5.	Brickwork and Blockwork	B5/SUM	
6.	Wood Works	B6/SUM	
7.	Ironmongery	B7/SUM	
8.	Steel and Metal Works	B8/SUM	
9.	Plastering and Paving	B9/SUM	
10.	Glazing	B10/SUM	
11.	Painting	B11/SUM	
12.	Provisional and Prime Cost Sums	B12/SUM	

Total Carried to Form of Tender

Signed: _____
(With Company Chop)

Date: _____

Figure 2.3 Example of a General Summary of the BQ

not part of the construction contract, and the unit rates should be used for the valuation of variations ordered by the architect.

The absence of quantities in the schedule of rates creates difficulties in tender assessment because different tenderers may use different quantities of the same item when calculating pricing. Therefore, the priced schedule of rates and related quantities will inevitably vary widely between tenderers. These variances make tender evaluation extremely difficult because there is no common basis for comparison. As a result, the SQR includes a schedule of rates with approximate quantities provided by the quantity surveyor, which are suitable for comparing tenders. These approximate quantities are not part of the construction contract and should not be subject to remeasurement during the final account stage. As the approximate quantities are given for reference only, tenderers may amend the approximate quantities and add any additional items to the SQR according to the drawings and specifications.

The format of the SQR is very similar to that of the BQ. For instance, the SQR can be presented in elemental or trade format. Preliminaries, preambles and provisional and PC sums are also included in the SQR. However, the item descriptions of the SQR are usually short and concise. Tenderers can also insert any additional items that they consider necessary for completing the contract at the end of each schedule.

2.5 Tender addendum

It is impractical for the design team to complete the entire design at the tendering stage. In addition, because of tight project schedules and high project complexity, mistakes and misunderstandings are bound to occur. It will be necessary to amend a BQ if tender drawings or specifications are revised, if clarifications are made for design development or if conflicts or discrepancies between tender documents are found after the tender is issued. A tender addendum may include additional or updated drawings and documents that are issued to tenderers before the tender closing date. It is thus a means of correcting or revising the tender documents and may involve inserting or removing any part of the tender documents that is erroneous or no longer applies. Tender addenda provide all tenderers with the opportunity to review the changes and respond appropriately.

Generally, if updates and changes are made to tender drawings or other tender documents, the quantity surveyor should issue a tender addendum to all tenderers. Every addendum page should be marked 'Addendum No. X' for identification, e.g., Addendum No. 1, 2 and so on. There is no limitation on the number of the tender addenda. However, every tender addendum creates additional tender workload and may result in abortive work for the tenderers. Because tendering schedules are often tight, tenderers may request extensions to the tendering period if an addendum has a substantial impact on tender pricing. Tenderers should incorporate all of the addendum pages into the tender documents according to the instructions for tender submission. Tenderers are requested to acknowledge the receipt of all addenda by signing an acknowledgement slip and returning it to the quantity surveyor. If any addendum pages are missing from a submitted tender, the quantity surveyor should issue a tender clarification (usually called a tender query) to request the outstanding pages.

2.6 Pre-tender estimate

A pre-tender estimate is prepared by the quantity surveyor, and it is an estimate of a bona fide tender before the date for tender returns. The quantity surveyor should estimate the likely cost of the tendered project based on the quantities in the BQ. However, the tender documents should not indicate the pre-tender estimate value. There are two key functions of the pre-tender estimate. First, it provides a means of reconciling the final cost plan with the tenders submitted by the tenderers. The quantity surveyor can assess the cost implication of any deviations identified in the reconciliation. If the pre-tender estimate exceeds the approved budget, the quantity surveyor should provide explanations to the employer and indicate necessary actions, such as cost-saving options, before the tender award. Second, the pre-tender estimate is a good basis for assessing and comparing returned tenders. As such, it is common practice to include the pre-tender estimate in the tender report to allow easy appraisal and comparison of the tenders received.

A pre-tender estimate is based on the full BQ including all tender addenda. When a pre-tender estimate is undertaken, it is important to ensure the reliability of the sources of cost data and to fully consider the critical factors that affect construction costs. The quantity surveyor's company's in-house cost database and the suppliers' quotations from the market are key sources of cost data used to prepare a pre-tender estimate. Moreover, a good pre-tender estimate must consider the form of the construction contract, the contract requirements, the current market conditions and any inherent and contingent project risk. Hence, the preparation of a pre-tender estimate is not a straightforward exercise, and its accuracy depends greatly on the skill and experience of the quantity surveyor.

2.7 Chapter summary

This chapter provided an overview of tendering and tender documentation. Tendering, a long-established practice for achieving true competition in the construction industry, is a process by which interested contractors are invited to submit tenders. The quantity surveyor plays a crucial role in achieving this goal by preparing tender documents that provide a common starting point for competitive tenders. BQ preparation is a critical task during this stage, and tremendous work is needed to determine quantities from the drawings and specifications provided by the design team. After the tender documents are sent to potential contractors, the quantity surveyor may need to prepare tender addenda that provide updated information to all tenderers before the tender closing date. Finally, a pre-tender estimate is prepared to provide a benchmark for assessing and evaluating the returned tenders.

2.8 Case study

Case one: post-tender addendum

Tender addenda are amendments to the tender documents issued by the quantity surveyor before the close of tendering. However, in the private sector, it is not uncommon for the quantity surveyor to issue post-tender addenda after tenders are returned. What are post-tender addenda?

Discussion

Generally, a tender addendum is used to update tender documents before tender return. A post-tender addendum is a change to the tender documents issued after the opening of tenders but before a construction contract is awarded. Such a change may be related to cost savings or errors that require a revised quote from all tenderers. As such, the tender sums offered by tenderers are subject to adjustments before the tender is awarded. The awarded contract sum is determined based on a tenderer's original tender sum and any adjustments arising from the post-tender addendum/addenda. Corrections of BQ errors and the inclusion of cost-saving options are some common adjustments.

Although a post-tender addendum is generally a means to obtain tenderers' revised prices for a contract, this arrangement may allow tenderers a second opportunity to submit a bid. For example, the second lowest tenderer might obtain unofficial information about the lowest tender's price. An addendum issued after the tender return would allow the second lowest tenderer to adjust his tender price to slightly less than the lowest tenderer's price with the consideration of the post-tender addendum. Hence, the quantity surveyor must always seek to be fair, open and transparent when advising the employer on the use of a post-tender addendum. Although rules differ in the private and public sectors, the principle of fairness always applies.

Case two: tender award criterion

The project budget is the employer's primary concern, so they use tendering to obtain competitive bids. To obtain low bids, the quantity surveyor may indicate in the tender documents that the construction contract will be awarded to the lowest bid. Is it appropriate to state this criterion in the tender documents?

Discussion

Generally speaking, an invitation to tender is not an offer. It does not commit the inviter to accept the lowest bid, the highest bid or, indeed, any bid. The call for tenders is usually an invitation to treat, and the resulting tenders are offers which the inviter is free to accept or reject.

This case can be considered from a contract law perspective. In a construction tender, it is common to find this statement: 'The Employer is not bound to accept the lowest or any tender which may be received'. The general rule for tender invitations is that it is an invitation to treat, not an offer. Hence, the employer is free to accept any tenderer's offer, whether it is the lowest bid or the highest. However, if the tender invitation states that the lowest bid will be accepted, the invitation is a unilateral offer that will be awarded to whomever complies with its terms (*Lobley Co Ltd v Tsang Yuk Kiu*, 1997). In fact, the employer should carefully examine highly competitive tender bids and should take the quality of the contracted work into account. In other words, they should enter into an agreement with a contractor who possesses the necessary technical skills, resources and financial backing. Hence, the employer should not accept an unreasonably low bid as this may involve risks that the project will not be completed at the required time, cost and quality standards. Thus, it is advisable not to state in the tender documents that the lowest bid will be accepted.

Case three: tender deposit

What is a tender deposit? Is it common to request a tender deposit from tenderers?

Discussion

A tender deposit is a deposit that a tenderer is required to pay at the time of tender submission in accordance with the tender documents. Submission of a tender deposit can be viewed as a pledge that the tender is made in good faith. A tender deposit is not a common requirement in construction tenders. If the employer requires a tender deposit, this requirement should be stated in the Conditions of Tender, which should include the amount of the tender deposit, the allowable methods of deposit payment (e.g., cheque and cashier's order) and the mechanism of deposit refund. There is no universal rule for calculating the amount of a tender deposit, but it should be a reasonable amount and not excessive. In addition, the quantity surveyor should state clearly in the tender documents that the deposit will be returned to tenderers without interest and indicate the criteria for returning the tender deposit.

Case four: special attendance

Tenderers are required to price attendance costs against PC sums in the BQ. However, in addition to general attendance, tenderers need to price special attendance for certain types of specialist work carried out by nominated sub-contractors. What is special attendance? What are the differences between general and special attendance provided by the contractor?

Discussion

The contractor should provide necessary site facilities for nominated sub-contractors and nominated suppliers to execute their work or supply materials. As such, the contractor should allow attendance based on the nature of the work and in cooperation with the nominated sub-contractors or nominated suppliers. Attendance should not be confused with general site preliminaries or the builder's work in connection with building services, which should be priced separately in the BQ.

There are two types of attendance: general and special. General attendance refers to the facilities established on site for the contractor's use and shared with the nominated sub-contractors and nominated suppliers. Examples of these site facilities include storage sheds, hoisting facilities, scaffolding, welfare facilities, safety provisions, rubbish clearance, and temporary power and water supplies. In some cases, the contractor is required to provide specific site facilities to certain nominated sub-contractors. These site facilities are the special attendance that these nominated sub-contractors need to execute tasks such as lift installation, window installation and structural steel installation. Examples of special attendance include special scaffolding (e.g., full-height scaffolding inside a lift shaft), special hoists (e.g., loading and unloading of chillers) and additional supplies of electricity (e.g., structural steel welding) and water (e.g., water-tight testing for curtain walls). To avoid confusion, it is important to state clearly in the tender documents which items are covered under general and special attendance and provided by the contractor and which items will be provided by the nominated sub-contractors and nominated suppliers.

Case five: lump sum contract variants

The two common types of lump sum contracts are lump sum contracts based on drawings and specifications and lump sum contracts based on BQ. What are the differences between them?

Discussion

All construction contracts allocate risks between the contracting parties. Unlike other contract types, such as remeasurement and cost reimbursement contracts, a lump sum contract is a contractual arrangement in which the employer transfers most of the cost of risks to the contractor. The contractor assumes a greater risk under a lump sum contract based on drawings and specifications than under a lump sum contract based on BQ because the SQR is used for pricing. In the private sector, the 2006 edition of the Agreement and Schedule of Conditions of Building Contract (without quantities) should be used for projects because of the characteristics of the SQR. According to clause 1.7 of the agreement, the unit rates contained in the SQR form part of the contract, and the contract sum should be based on the contract drawings or described in the specifications.

Both the BQ and SQR can be used to obtain prices from contractors and for the valuation of variations. The choice of the BQ or SQR depends on the project nature and the chosen procurement route. The SQR is commonly adopted for small works and projects where it is not possible to prepare a BQ at the time of tendering. The reason most likely is that the progressive design is sufficient for tendering purposes but not firm enough to provide accurate quantities. The SQR may also simplify the tender preparation process because quantities are not measured by the quantity surveyor in detail. However, the SQR cannot be misused by the quantity surveyor who wants to take the advantages of savings in tender production time and error-free quantity measurement.

3 Tender examination and contract award

Learning outcomes

Upon completion of this chapter, you should be able to do the following.

1   Understand the purposes of tender examination and analysis.
2   Know the process of analysing returned tenders.
3   Understand the practice of contract binding.
4   Be able to examine tenders and award a contract.

3.1 Introduction

A construction tender is a submission made by a prospective contractor in response to a tender invitation. Once tenders have been received, an assessment process must be undertaken to select a capable contractor who will carry out the construction work for the tendered project. This process is known as tender examination. Typically, the tenderer who offers the lowest bid is awarded the construction contract. However, using a single selection criterion may not result in the selection of the best tender. To ensure that the construction works are completed to the required time, cost and quality standards, the employer should enter into a contract with a capable contractor who possesses the necessary technical skills, resources and financial backing. This chapter examines the tender analysis process in detail and describes the key duties of the quantity surveyor in tender evaluation. It then examines the tender award and contract binding. Finally, some cases are discussed.

3.2 Tender analysis

After tenders are returned, the project team members engage in tender analysis, following the required procedures for opening and recording tenders. Once the tenders are opened, the architect and engineers are responsible for assessing their technical aspects, and the quantity surveyor assesses the cost and other contractual issues. During the tender analysis, the project team members often correspond with tenderers to obtain clarification and confirmation, such as regarding the withdrawal of tender qualifications.

Tender interviews are held before the contract is awarded. The interviews not only allow for clarification of matters that may otherwise lead to disputes but also give the employer insights into potential problems with the tenders. The quantity surveyor should consider all of the tenderers' replies and written responses obtained from the tender correspondence and arising from the interviews. Finally, the quantity surveyor prepares a tender report, which

DOI: 10.1201/9781003212355-3

includes an analysis of the lowest three or four tender submissions (if only tender prices are being scored) and makes a recommendation.

3.2.1 Tender opening

When the tenders are returned, they should be opened in the presence of the project team members, including the employer's representatives, the architect, the engineers and the quantity surveyor. The tender submission deadline should be strictly upheld, and late tenders should not be accepted. During the tender opening meeting, the quantity surveyor should record the tender sums offered by each tenderer on the tenders received form. A sample form is displayed in Figure 3.1.

The tender sum is the total tendered amount shown in the Form of Tender. If an alternative tender is applicable, the tender sums in both Forms of Tender (i.e., Tenders A and B) should be recorded. After ranking the tenderers in ascending order by tender sum, the tenders received form should be signed by all parties present at the meeting as witnesses of the tender opening process. The tender results should not be disclosed, and all parties at the meeting should be reminded of their confidentiality. After tenders are opened, the project team members should reach a consensus with the employer on the number of tenders to be shortlisted for detailed evaluation.

CONFIDENTIAL

TENDERS RECEIVED FORM FOR
Proposed Residential Development

DATE
24 July 2018

Names of Tenderers (in alphabetical order)	Tender A		Tender B*		
	Tender sum HK$	Ranking	Tender sum HK$	Proposed contract period	Ranking
Chan Kei	376,104,300	5			
Chun Wing	329,080,060	2			
Dai Lik Kee	354,490,000	4			
Eddy City	310,500,200	1			
Finlon Tech	346,500,000	3			
Gamming	428,110,000	7			
New Tak	398,360,000	6			

Opened and witnessed by:

_____ (company) _____ (name) _____ (signature)
_____ (company) _____ (name) _____ (signature)
_____ (company) _____ (name) _____ (signature)

* Tender B is an alternative tender for the tenderer's proposed contract period and is not applicable in this case

---

Figure 3.1 Example of a tenders received form

3.2.2 Technical check

The received tenders are subject to scrutiny and checking by the quantity surveyor. Both the front and pricing parts (e.g., the BQ) should be checked for irregularities. Some common irregularities are as follows.

1   Failure to complete and/or sign the Form of Tender.
2   Failure to sign the General Summary of the BQ.
3   Non-submission of required documents stated in the Conditions of Tender.
4   Failure to incorporate amendment pages arising from tender addenda.
5   Missing pages in tender documents.
6   Amendment of BQ descriptions and/or insertion of qualifying notes by tenderers in tender documents.

Items 1 to 5 above relate to incomplete submissions and should be treated as general errors. Item 6 should be dealt with as a tender qualification issue. Such proposals should be treated as counteroffers that do not meet the original tender requirements.

3.2.3 Examination of the BQ prices

The BQ prices should be checked for arithmetical errors and pricing anomalies. Arithmetical errors are those made by tenderers in the pricing of the BQ. Arithmetical errors should be rectified as described in the Conditions of Tender. The tender sum should remain unchanged. Some typical arithmetical errors are as follows.

1   Errors in multiplying the quantity by the unit rate (e.g., quantity x unit rate).
2   Errors in casting the cash column (e.g., summation of item amounts on a single BQ page).
3   Errors in carrying forward a total amount to the collection/summary page (e.g., transferring the total amount from an individual BQ page to a Bill summary page or from a Bill summary page to the General Summary page).
4· Errors in casting a collection/summary page (e.g., the summation of page amounts on a Bill summary page or of bill amounts on the General Summary page).
5   Errors in entering the unit rate.

Pricing anomalies may include inconsistency, missing prices, group pricing and unreasonable price levels. The quantity surveyor should read through the entire BQ and inform tenderers about their pricing anomalies. Some common examples of pricing anomalies are as follows.

1   Unreasonably high or low unit rates.
2   Marking BQ items as 'excluded'.
3   Marking BQ items with a 'dash' or omitting pricing.
4   Pricing identical items differently, whether in the same bill section or separate bill sections.
5   The bill of preliminaries is not priced in detail.

The quantity surveyor should propose actions for tenderers to take concerning any pricing anomalies.

For item 1, reasonable rates are usually proposed to tenderers for the valuation of future variations after the contract is awarded.

For item 2, marking a BQ item as 'excluded' is treated as a qualification of the tender. In such cases, tenderers should be asked to withdraw the qualification and to price the concerned item. The additional amount introduced should then be treated as an arithmetical error. Items 3 and 4 that give rise to errors after re-pricing should receive the same treatment.

For item 5, tenderers should be asked to submit cost breakdowns of the preliminaries bill or a group of items on the preliminaries bill. However, if any items on the preliminaries bill are priced abnormally high, the quantity surveyor should address them at the project outset to avoid making substantial payments.

In addition, if a tender is front-loaded to improve the contractor's cash flow during the early project stages, this should be reported to the employer.

3.2.4 Three-rate bills

The quantity surveyor is responsible for analysing the reasonableness of the tender prices and usually prepares three-rate bills for this purpose. Generally, the lowest tenderer's price will form the basis of price comparison. The prices offered by the second and third lowest tenderers are then compared with the lowest tenderer's price to obtain a reasonable price trend. The quantity surveyor should notify the shortlisted tenderers if unreasonably high or low rates are observed with the consideration of their bills of preliminaries priced separately. Three-rate bills also establish an important database for cost estimates in future projects. The format of a three-rate bill is illustrated in Table 3.1.

Table 3.1 Example of a three-rate bill

	BQ Description	Qty.	Unit	Rate (HK$/m²)		
A	25mm thick cement and sand (1:3) paving	480	m²	150	160	175
				(priced by the lowest tenderer)	(priced by the 2nd lowest tenderer)	(priced by the 3rd lowest tenderer)

3.2.5 Tender qualifications

Tender qualifications mean tenderers propose some terms or conditions that contradict the original tender requirements. For instance, a tenderer might propose an amendment to the payment terms or the liquidated and ascertained damages rate stipulated in the tender documents. Generally, tender qualifications are not accepted, and tenderers should withdraw them upon the request of the quantity surveyor. For fairness, a tender should be rejected if a qualification is not withdrawn upon request.

There is no difference between the private and public sectors in the treatment of tender qualifications. However, there are some cases in which qualifying tenders may be considered in private-sector projects. For example, some tenderers may propose economic solutions that are advantageous to the employer, such as alternative construction materials. In this case, the quantity surveyor should bring these alternative proposals to the design team for review and to the employer for consideration. The quantity surveyor should provide a professional assessment of the financial effect of the alternative proposals and make a fair comparison between all tenderers.

However, if cost is the employer's critical concern, the quantity surveyor should ask the design team to investigate any alternative options at the design stage. In this case, the provision for pricing design alternatives can be incorporated into the tender documents in advance to obviate the possibility of 'advantageous qualifications' during the tendering process.

3.2.6 Tender query

If irregularities are found in the tenders returned, the quantity surveyor should issue tender queries (also called tender clarifications) to request clarifications. There are three general types of tender queries. The first type concerns outstanding submissions. Tenderers are required to submit outstanding information for the tender assessment, such as an outline of the construction programme, a proposed organisation chart, a proposed plant schedule or a detailed safety supervision plan. If the tenderer fails to submit the required information with the tender, the quantity surveyor should send a tender query. However, if the outstanding submitted documents are to be included in the calculation of the total tender score (e.g., technical proposals must be submitted with the tender under the two-envelope tender system), then the submission of those outstanding documents are not allowed.

The second type of tender query concerns the notification of errors or anomalies found in the BQ. If a tender includes arithmetical errors or pricing anomalies, the quantity surveyor should prepare a tender query that addresses them for the tenderer's reference. The concerned tenderer is required to acknowledge the identified errors or anomalies and to agree to abide by the submitted tender sum. The tenderer is also required to price the excluded BQ items or to come to agreement with the quantity surveyor on reasonable rates proposed for the valuation of future variations after the contract is awarded.

The third type of tender query concerns qualifications proposed by tenderers. Generally, qualifications that contradict the original tender requirements are not accepted. The concerned tenderers are required to withdraw the qualifications stated in the tender query and to agree to abide by the tender sum. The types of tender query and the tenderers' corresponding actions are summarised in Table 3.2.

3.2.7 Tender interview

A tender interview is a formal meeting in which the project team meets with the tenderers. Such interviews are particularly useful for the project team to obtain further details about the tenders submitted. A long list of interviewees is usually not recommended, and typical practice is to invite the shortlisted tenderers only. The project team should also pay attention to scheduling the order of interviewees to avoid exposing the tenderers' rankings at the interview venue.

During a tender interview, the design team members focus on technical issues and the quantity surveyor asks questions about financial and contractual issues. To gather necessary information, the project team should determine beforehand what problems are to be addressed by each member. For example, a tender interview is useful as a follow-up to tender queries such as outstanding replies or further clarifications. A project team member should record all verbal replies from tenderers, and the quantity surveyor should pay close attention

Table 3.2 Examples of tender queries and tenderers' actions

Tender Query Subjects	Examples of Irregularities	Examples of Tenderer's Actions
Outstanding Submission	Failure to submit the information required in tender documents	Submit the outstanding information
BQ Pricing	Arithmetical errors, pricing anomalies, etc.	<u>Arithmetical errors</u> Acknowledge the errors and abide by the tender sum <u>Unreasonably high unit rates</u> Agree to reasonable rates for the purpose of valuation of variations <u>Marking BQ items as 'excluded'</u> Withdraw the exclusions, price the excluded BQ items and abide by the tender sum
Qualifications	Proposed terms or conditions to qualify the tender	Withdraw all qualifications that contradict the tender documents and abide by the tender sum

to statements that have cost and contractual implications. All discussed items should be confirmed in writing by tenderers after the interview. Attention should be paid to written replies as they generally have contractual effects and become contractually binding.

3.2.8 Extension of the tender validity period

All of the tenders returned remain binding until the expiry of the tender validity period, which is stipulated in the Form of Tender. Normally, the period is 90 days from the tender closing date, but it can be any period requested by the employer. No tender can be accepted after that period. If the quantity surveyor anticipates that an extension will be required, a letter should be issued to all tenderers asking for confirmation that they are willing to extend the tender validity period without changing their tender sums. Tenderers should be reminded that such a letter does not signify acceptance or rejection of their tenders. If a tenderer refuses to extend the tender validity period upon written request by the quantity surveyor, his tender will be ineligible for acceptance after the original tender validity period expires. Furthermore, a tenderer's agreement to extend conditionally would render his tender a qualified bid, which may still be rejected. All letters and written replies related to the extension of the tender validity period should be included in the tender report.

3.3 Tender report

The purpose of a tender report is to advise the employer of the suitability of a particular tender to be the basis for a construction contract. The report summarises the results of the tender analysis of the shortlisted tenders. The quantity surveyor should submit the report to the employer as soon as possible, taking into account the project commencement date, period of acceptance of the tender and any dates on which the employer may have internal management meetings. The tender report should cover the following key items and the actions taken in regard to them.

1 The expiry date of the tender validity period.
2 Any arithmetical error of each shortlisted tender found.
3 The general tender price level and any irregular pricing, such as front-loading of tender prices.
4 Errors and inconsistencies in pricing and unreasonable rates that may have significant effects on future variation costs.
5 Any qualifications proposed by the shortlisted tenderers together with their details and actions taken by the project team.
6 A comparison of the lowest tenderer's price and the pre-tender estimate to determine whether the price is reasonable.
7 Recommendations to the employer regarding whether the shortlisted tenders are contractually suitable for acceptance or alternative arrangements are required, such as re-tendering.

The report should include an executive summary that highlights the comparison of the tender sums. Other information, such as the tender validity period, addendum notices, arithmetical checks and critical comments on tenderers should also be included. Appendices that provide in-depth tender evaluations should be attached to the report for the employer's reference. In general, a tender report contains the following appendices:

1 submission status of tenderers;
2 comparative bill-by-bill breakdown of tenders;
3 comparison of major unit rates; and
4 tender correspondence.

3.3.1 Submission status

The submission status appendix lists the information required from tenderers and the submission status of each shortlisted tenderer. The information includes all of the submission items listed in the Conditions of Tender, acknowledgements of tender addenda and any schedules for equipment and delivery of building services installation works. The appendix serves as a checklist, and failure to submit the required information may cause the tender to be disqualified.

3.3.2 Comparative bill-by-bill breakdown

The comparative bill-by-bill breakdown appendix shows the BQ breakdown of the shortlisted tenderers, making it easy to compare bill amounts. If any tenderers commit arithmetical errors in their BQ, the bill amounts should reflect the corrected amounts. As such, the comparative bill-by-bill breakdown shows the corrected total sum of each tenderer and the aggregate amounts of the arithmetical errors, if any. These amounts form the basis for calculating the adjustment percentage according to the error correction rule as shown in the

Conditions of Tender. The calculation and adoption of the adjustment percentage is further explained in the case study below. It is good practice for the quantity surveyor to present the reconciliation of the pre-tender estimate with the shortlisted tenders in the bill-by-bill breakdown, as it ensures that the tenders are easily compared with one another and with the pre-tender estimate. Any anomalies in the bill amounts of the shortlisted tenderers can thus be identified. Figure 3.2 shows an example of a comparative bill-by-bill breakdown prepared by a quantity surveyor.

Tender Report – Appendix B
Proposed Residential Development
Comparative bill-by-bill breakdown of tenders

	1st Lowest Tender	2nd Lowest Tender	3rd Lowest Tender	Pre-tender Estimate
	HK$	HK$	HK$	HK$
Bill No. 1 – Preliminaries	61,523,000.00	64,840,000.00	63,569,000.00	63,142,000.00
Bill No. 2 – Preambles	-	-	-	-
Bill No. 3 – Excavation	36,564,000.00	37,295,000.00	37,661,000.00	36,930,000.00
Bill No. 4 – Concrete Works	104,515,000.00	106,605,000.00	107,650,000.00	105,560,000.00
Bill No. 5 – Brickwork and Blockwork	13,128,000.00	13,391,000.00	13,522,000.00	13,260,000.00
Bill No. 6 – Wood Works	38,791,000.00	39,567,000.00	39,955,000.00	39,174,000.00
Bill No. 7 – Ironmongery	8,253,000.00	8,418,000.00	8,501,000.00	8,335,000.00
Bill No. 8 – Steel and Metal Works	10,341,000.00	10,548,000.00	10,651,000.00	10,445,000.00
Bill No. 9 – Plastering and Paving	53,694,000.00	54,768,000.00	55,305,000.00	54,231,000.00
Bill No. 10 – Glazing	4,592,000.00	4,684,000.00	4,729,000.00	4,638,000.00
Bill No. 11 – Painting	16,128,000.00	16,456,000.00	16,612,000.00	16,285,000.00
Bill No. 12 – Provisional and PC Sums	148,210,000.00	148,570,000.00	149,156,000.00	148,000,000.00
Corrected Total Sum	495,739,000.00	505,142,000.00	507,311,000.00	500,000,000.00
Arithmetical Errors	-	(32,000.00)	744,000.00	-
Tender Sum (as Form of Tender)	495,739,000.00	505,110,000.00	508,055,000.00	500,000,000.00
% Higher Than the 1st Lowest Tender	–	+1.89%	+2.48%	+0.86%

Figure 3.2 Example of a comparative bill-by-bill breakdown

3.3.3 *Comparison of major unit rates*

This comparison presents the unit rates priced by the shortlisted tenderers for major BQ items. Generally, the unit rates that have significant cost impacts on a construction contract are considered major rates in the BQ. Unit rates priced for costly BQ items or substantial quantities in the BQ are examples of major rates. Comparison can reveal any unreasonable rates offered by the shortlisted tenderers. The quantity surveyor should pay attention to the BQ quantities of overpriced items, particularly quantities that are provisional or under-measured. A reminder to the design team should be included in the tender report regarding the cost implications of major variations in overpriced items ordered at the construction stage. For private projects, it is general practice for the quantity surveyor to propose reasonable rates for future valuation of variations to the concerned tender via tender query.

Tender Report – Appendix C
Proposed Residential Development
Comparison of major unit rates

BQ Ref.	Description	1st Lowest Tender	2nd Lowest Tender	3rd Lowest Tender	
3/1/A	Excavating basement ≤ 2.00m deep	$230.00/m³	$240.00/m³	$500.00/m³	*
3/2/D	Excavating for footings ≤ 1.50m deep	$210.00/m³	$220.00/m³	$230.00/m³	
3/3/B	Remove excavated materials off site	$300.00/m³	$310.00/m³	$320.00/m³	
4/1/D	Reinforced concrete grade 30	$1,100.00/m³	$1,200.00/m³	$1,300.00/m³	
4/3/B	Mild steel rod reinforcement	$9.00/kg	$9.50/kg	$9.80/kg	
4/4/D	High tensile rod reinforcement	$9.00/kg	$9.50/kg	$9.80/kg	
4/5/D	Sawn formwork to soffits of suspended slabs	$400.00/m²	$600.00/m² *	$420.00/m²	
4/5/E	Sawn formwork to columns and walls	$400.00/m²	$600.00/m² *	$420.00/m²	
5/1/C	105mm solid concrete block walls	$300.00/m²	$320.00/m²	$340.00/m²	
6/4/D	Timber skirting	$120.00/m	$130.00/m	$150.00/m	
8/1/B	Structural steelwork	$30.00/kg	$31.00/kg	$32.00/kg	
8/6/E	Galvanised mild steel in railings and framed work	$31.00/kg	$32.00/kg	$60.00/kg	*
8/8/F	Metal panel suspended ceiling	$650.00/m²	$670.00/m²	$700.00/m²	
9/1/C	Two coats internal lime plaster to soffit and beams	$100.00/m²	$110.00/m²	$120.00/m²	
9/2/D	25mm cement and sand (1:3) paving	$150.00/m²	$160.00/m²	$175.00/m²	
9/5/A	Ceramic tiles bedded to floor screed	$400.00/m²	$420.00/m²	$600.00/m²	*
11/1/D	Two coats of emulsion paint to plastered wall	$100.00/m²	$110.00/m²	$120.00/m²	

* *High unit rate*

Figure 3.3 Example of a comparison of major unit rates

3.3.4 *Tender correspondence*

Tender correspondence includes the letters and documents exchanged between tenderers and the project team members related to the tenders. They contain essential information such as tender submissions, tender queries and replies, and written confirmations of issues verbally discussed in the tender interviews. The items in the tender correspondence of the successful tenderer that have contractual effect become part of the contract correspondence in the contractual binding process.

3.4 Award of a tender

After reviewing the tender reports prepared by the consultant team, the employer decides whether to award the construction contract to the recommended tenderer. A contract award means that a tenderer is the contractor for the tendered construction project. Hence, a set of contract documents should be prepared for contract execution by the employer and the contractor. It is relatively common for a construction contract to remain unsigned until well after the contract work has begun. Contract execution is important because a signed contract indicates that the parties have agreed to their contractual duties and obligations. Contract documents include the contract drawings, the articles of agreement and conditions, and the front and pricing parts of the tender documents. A large volume of contract documents are usually produced for contract execution. Hence, sufficient time should be allowed for the quantity surveyor to prepare the documents and bind them in correct order. Under tight time constraints, a letter of intent can be used to enable the contractor to begin construction work prior to the formal execution of the construction contract.

3.4.1 *Letter of intent*

The purpose of a letter of intent is to allow the contractor to begin the construction work before the formal execution of the construction contract. Thus, a confirmation that promises reasonable reimbursement in the absence of an enforceable agreement is essential. A letter of intent serves this purpose, as it expresses an intention to enter into a construction contract in the future. Generally, a letter of intent creates no contractual relationship until a future construction contract has been entered into. However, several legal cases support the contractual significance of a letter of intent, as courts have held that it is a binding contract between parties. To prevent unnecessary disputes, it is advisable to avoid using a letter of intent, as it does not cover all of the eventualities stipulated in a construction contract. In other words, a letter of intent offers commercial convenience for project commencement but should never be considered an alternative to a full contract.

3.4.2 *Letter of award*

Valid acceptance of a tender must be properly communicated to the offeror. Commonly, a letter of award is drafted by the quantity surveyor to confirm the employer's acceptance of a winning tender. The typical content of a letter of award includes the contract sum, the contract period, the rate adjustment for an arithmetical error or a commercial discount (if any), any documents forming part of the construction contract, the contract correspondence and other requirements with contractual effects confirmed via tender queries or during the tender interview. The contract award and formal contract execution may not necessarily occur at the same time. Hence, a letter of award is a written confirmation that the employer has awarded a construction contract to a successful tenderer before the execution of a formal agreement. However, to avoid ambiguity, the quantity surveyor should clearly state that the letter of award, confirmed by the contractor returning a signed copy of such a letter, constitutes a binding agreement between the employer and the contractor.

3.4.3 *Contract documents and execution*

Contract binding is essential for the execution of a formal contract. A construction contract is not executed until both parties have signed and witnessed it, and the quantity surveyor often plays a critical role in assembling the contract documents and preparing them for the contract parties to sign. Generally, the tender documents form an integral part of the contract documents, which include the Conditions of Tender, Form of Tender, General Conditions of Contract (GCC), Special Conditions of Contract (SCC), Contract Correspondence, Specification, BQ (or SQR) and contract drawings. In contract execution, a contract booklet is coupled with a set of contract drawings prepared by the design team.

Contract binding involves several key practices.

1 The Articles of Agreement should be completed by filling the necessary information, such as the names of the employer and the contractor, the project title and location, the awarded contract sum and the name of the architect and the quantity surveyor.
2 The GCC bound into the contract documents should reflect any amendments of the contract clauses by SCC. For example, a remark should be stamped against the clauses amended by SCC, and the non-applicable clauses should be crossed out.
3 If the tender sum differs from the awarded contract sum, a remark should be added in the Form of Tender where the original tender sum is shown. In addition, a build-up for the awarded contract sum based on the original tender sum with the adjustment of other agreed-upon amounts should be clearly indicated in the letter of award.
4 An adjustment factor arising from an arithmetical error or a commercial discount should be stamped on every BQ page to which a rate adjustment of the BQ items for future variations is applicable.
5 Correspondence with contractual effect that forms part of the construction contract should be included and bound as contract correspondence.
6 If the contract documents contain a great number of pages and a second volume is needed, a separate signing page should be prepared and incorporated into the second volume.

The quantity surveyor should collate the contract documents and arrange for copying and binding. Generally, two sets of the contract documents must be signed by the contract parties: the contract original and the contract duplicate. After contract execution, the employer retains the contract original, and the contract duplicate is issued to the contractor. The need for extra certified true copies depends on the employer's and the project team's requirements. Once the contract documents, including the contract drawings, are ready for execution, both parties can sign the page of attestation in the Articles of Agreement of the GCC to confirm their intention to execute the terms and conditions of the construction contract.

A construction contract can be executed under seal or under hand. A contract under seal is executed as a deed that must be signed by the person who is to be bound, and the signature must be attested by a witness. The witness should sign and provide his/her name and occupation. The limitation period for a contract under seal is 12 years. In contrast, a contract under hand is simply signed by the contract parties and has a limitation period of six years. If necessary, the quantity surveyor should check the Contractor's Articles of Memorandum and Articles of Association to ensure that the requirements on sealing and signing the construction contract are in accordance with their relevant provisions on use of the seal and the person(s) authorised to sign contracts for the company.

3.5 Chapter summary

This chapter provides some guidance on quantity surveying practice following tender return. With input from the project team members, the quantity surveyor should provide professional advice on the returned tenders and summarise the results in a tender report. The tender report provides an overview of the shortlisted tenderers and lists all of the technically and contractually cleared issues for the employer's consideration on awarding the construction contract. The winning tenderer becomes the contractor of the tendered project. The quantity surveyor is then responsible for preparing the contract documents, which are an important deliverable for contract execution, whereby the contract parties agree to the terms, contractual duties and obligations stipulated in the construction contract.

3.6 Case study

Case One: *withdrawal of the tender offer before the employer's acceptance*

You are the quantity surveyor for a residential project. During tender analysis of the superstructure of the main contract, you are informed that one of the tenderers, Finlon Tech Engineering Company Limited, has withdrawn its tender in writing without providing a reason. This is a conforming tender that includes no significant errors. According to the Form of Tender, the tender validity period has not yet expired, and all tenders can still be accepted. Advise the employer regarding this case.

Discussion

Based on the general rule of contract law, an offer may be withdrawn at any time before it is accepted. Hence, it seems that Finlon Tech is free to withdraw its tender at any time before it is accepted by the employer. However, under the current practice, the Form of Tender usually includes the following statement:

> I/We agree to abide by this tender for a period of 90 days commencing from and including the day following the date fixed for receiving it and it shall remain binding upon me/us and may be accepted at any time before the expiration of that period.

This term becomes an implied contract that binds every tenderer to keep his/her tender open for acceptance for 90 days from the tender closing date. The employer considers all timely and conforming tenders. Thus, when the tender documents require tenderers not to amend or withdraw the tender for a specified time, tenderers are contractually obliged to comply, and the employer is entitled to seek damages for a breach if a tenderer fails to do so (*City Polytechnic v Blue Cross*, 1993).

Case two: the tender sum in the Form of Tender is incorrect

During the tender examination of a residential project, you find a discrepancy in the tender sum in the tender submitted by Chan Kei Construction Company Limited. The tender sums stated in the Form of Tender in numbers and words are 'HK$376,104,300' and 'HK Dollar Three Hundred and Seventy-six Million One Hundred Thousand Only', respectively. You check the General Summary of the BQ and find that the total bill amount is HK$376,104,300. As the quantity surveyor for this project, how would you advise the employer regarding this discrepancy?

Discussion

In the Form of Tender, the amount written out in words must be equal to the amount written in numbers. If the two differ, the amount in numbers (i.e., HK$376,104,300) takes precedence because it agrees with the total amount in the General Summary of the BQ. The Technical Circular (Works) No. 41/2002 published by the Hong Kong Government provides further explanation (Environment, Transport and Works Bureau, 2002). It states that if neither the tender sum in numbers nor the sum in words stated in the Form of Tender agrees with the total amount stated in the General Summary of the BQ, the amount in numbers should be taken as the tender sum. If the amount in either words or numbers is blank or illegible, the remaining amount should be taken as the tender sum. If the amounts in words and numbers are both blank or illegible, the tender is invalid.

Case three: arithmetical error in the BQ

You are the quantity surveyor for a residential project that is in the tender return stage. Your assistant finds that Eddy City Construction Limited has committed an arithmetical error in the BQ. However, your assistant has no idea how to rectify the error. Advise your assistant on where to find the error correction rule for the tender documents and how the error should be rectified and adjusted according to the rule.

Discussion

There is no universal approach to dealing with arithmetical errors. In common practice in Hong Kong, the correction rule is usually stated in the Conditions of Tender. However, the tender sum should remain unchanged after the error is rectified and adjusted. Under this principle, the difference between the corrected total sum and the tender sum represents the net aggregate amount of an arithmetical error, which can be a positive or negative value. As the error will be calculated as a percentage adjustment (negative or positive percent) applied to all unit rates/sums in the BQ, any items not subject to such adjustment should be excluded from the calculation, such as the preliminaries and other pre-priced items such as the provisional sums, contingencies, prime cost sums and amounts of prime cost rate items. The percentage can be determined using the following example formula:

$$\text{Percentage} = \frac{\text{Net aggregate amount of arithmetical errors } (+/-)}{\begin{array}{l}\text{Corrected total sum} - \text{Preliminaries} - \text{Provisional sums (incl.}\\ \text{contingencies)} - \text{PC sums (excl. profit \& attendance)} - \text{Amounts}\\ \text{of PC rate items}\end{array}} \times 100\%$$

After the percentage adjustment is calculated, the revised rates/sums are used for the valuation of variations ordered by the architect and the valuation of interim payment. However, the adjustment always excludes preliminaries, provisional sums, contingencies, prime cost sums and prime cost rate items. If the calculated percentage is minor, the net aggregate of the arithmetical error will be adjusted by one figure in the General Summary of the BQ, and no percentage adjustment will be applied to the BQ rates, as mentioned. It is common practice to define a percentage that is minor in nature as ± 0.25%.

Case four: contradiction between the letter of award and the contract appendix

After the construction contract is executed by the employer and the contractor, the architect finds that the contract period shown in the Letter of Award is not in line with the number of days between the commencement date and completion date given in the contract appendix. As the quantity surveyor, what is your advice to the architect regarding this discrepancy?

Discussion

According to the 2005 edition of the Agreement and Schedule of Conditions of Building Contract published by the Hong Kong Institute of Architects, the Hong Kong Institute of Construction Managers and the Hong Kong Institute of Surveyors, the order of precedence of the documents forming the contract should follow the order of documents listed in clause 5.1:

1   The Articles of Agreement and the Appendix;
2   The Form of Tender submitted by the contractor together with the employer's letter of acceptance of the contractor's tender and any correspondence between the parties expressed to form part of the contract;
3   The Special Conditions of the contract, if any;
4   The General Conditions of the contract; and
5   The Contract Drawings, the Specification and the Contract Bills.

Thus, the contract appendix should prevail in this case, and the contract period should be determined based on the commencement date and completion date stated in the contract appendix.

4 Interim valuation and payment

Learning Outcomes

Upon completion of this chapter, you should be able to do the following.

1 Understand the principle of interim valuation.
2 Understand the contract provisions of interim valuation.
3 Know the time frame and mechanism of interim valuation and payment.
4 Prepare an interim valuation.

4.1 Introduction

Under a construction contract, the contractor is obliged to build, and the employer is obligated to pay the contractor. Thus, the employer is contractually liable for paying the contractor. The interim valuation is prepared by the quantity surveyor, who determines the amount of payment indicated on an interim certificate. The architect is contractually bound to issue interim certificates during the period stated in the contract. Interim valuation is a complex process that involves multiple project members. The employer is the payer, the contractor is the payee, the architect is the certifier and the quantity surveyor is the valuer. Hence, all of them play significant roles in successful interim payments. This chapter first introduces the contract clauses relevant to interim valuation and certificates. The payment time frame and mechanism based on these contract provisions are then discussed in detail. Under-certification places unreasonable financial burden on the contractor, whereas over-certification can create risk to the employer. This chapter therefore also discusses valuation practice for quantity surveyors preparing interim valuations.

4.2 Contract provisions

The 2005 edition of the Agreement and Schedule of Conditions of Building Contract published by the Hong Kong Institute of Architects, the Hong Kong Institute of Construction Managers and the Hong Kong Institute of Surveyors ('the Contract') has several clauses that set out the way in which the contract sum, as adjusted under various provisions in the Contract, is to be paid to the contractor and its nominated sub-contractors and nominated suppliers. Some clauses specify the payment time frame and others address the interim valuation and payment mechanism requirements. Table 4.1 summarises the clauses abstracted from the Contract related to interim valuation and payment. The contract provisions, key parties and purposes are also listed.

DOI: 10.1201/9781003212355-4

Table 4.1 Interim valuation and payment clauses under the Contract

Clause	Provisions	Key Parties Involved	Payment Time Frame/Mechanism
32.1 Interim Certificates and Interim Valuations	(1) Period of issue interim certificates (2) Period for payment of certificates (3) Issue interim certificates after substantial completion (4) Payment application (5) Valuation of interim payment (6) Decision on payment deduction	Employer Architect Quantity Surveyor Contractor	Time frame Time frame Mechanism Time frame Time frame Mechanism
32.2 Estimate of Amount Due in Interim Certificate	(1) Amount due in an interim certificate (2) Estimated gross valuation (3) Estimated amounts included in gross valuation (4) Estimated set-off amounts included in gross valuation	Quantity Surveyor Contractor	Mechanism Mechanism Mechanism Mechanism
32.3 Off-site Materials or Goods	Decision on payment for off-site materials or goods	Architect Quantity Surveyor Contractor	Mechanism
32.4 Calculation of Retention	(1) Payable amounts excluded from retention calculation (2) Limit of retention (3) Separate retention calculation for the main contractor and the nominated subcontractors/suppliers	Quantity Surveyor Contractor	Mechanism Mechanism Mechanism
32.5 Retention Rules	(1) Retention held upon by trust subject to the employer's rights (2) Release of retention (3) The amount of retention to be held for a section or relevant part (4) Release of retention	Employer Architect Quantity Surveyor Contractor	Mechanism Mechanism Mechanism Mechanism
32.13 Late Payment	Interest for late payment by the employer	Employer Quantity Surveyor Contractor	Mechanism

The quantity surveyor should also take into account the Security of Payment Legislation (SOPL) proposed by the Hong Kong Government. The Development Bureau issued the SOPL consultation document in June 2015 to consult the public on regulation of certain aspects of the payment practice in the construction industry (DEVB, 2015). At the time of publication of this book, the Legislative Council had not passed the SOPL, so it is not yet in effect. According to the consultation document, the SOPL will have a profound impact on the contract provisions related to interim valuation and payment, such as by prohibiting the 'pay when paid' provision, providing the right to adjudication for dispute resolution and the right to suspend, and reducing the rate of work progress.

4.3 Payment time frame

The Contract specifies the time frame of interim valuation and payment certificates for all parties. The periods specified under the contract conditions are in calendar days that include Sundays and other holidays according to the General Holidays Ordinance of Hong Kong. Under all circumstances, the quantity surveyor should note any relevant departures from the Contract by referring to the Special Conditions of the Contract.

4.3.1 *Payment application and payment certificates*

In general, interim valuation can be undertaken periodically or in stages depending on the payment terms agreed upon by the contract parties. If the Contract specifies regular progress payments to the contractor, the quantity surveyor should check the payment period prescribed in the Contract. The period of interim certificates is stipulated in the Contract appendix. Interim certificates are issued by the architect at calendar monthly intervals before the Substantial Completion Certificate is issued if monthly progress payment has been chosen. In practice, the quantity surveyor and the contractor should agree on the time of each interim payment immediately after the project commences, with the first interim certificate issued no later than 42 days after the commencement date (clause 32.1(1)).

Interim valuation is based on a detailed cost breakdown that is generally prepared by the contractor. The cost breakdown constitutes a statement of the estimated gross valuation of the work in progress. The contractor's submission of the statement should be made at least 14 days before the interim certificate is due to be issued by the architect (clause 32.1(4)). To determine the amount due in interim certificates, the architect should have the interim valuations verified by the quantity surveyor. Hence, the quantity surveyor is responsible for assessing the contractor's payment application by visiting the site and checking that the work has been properly carried out and that materials/goods have been delivered to the site. The quantity surveyor should submit the valuation to the architect at least seven days before the interim certificate is due to be issued (clause 32.1(5)). When an interim certificate has been issued to the contractor, the employer is given a grace period in which to settle payment, which must be made within 14 days of the date of the interim certificate, or such other time as stipulated in the Contract appendix (clause 32.1(2)).

The interim valuation covers the work completed and the materials/goods delivered, including the work carried out by the nominated sub-contractors and the materials/goods supplied by the nominated suppliers. Although the contractor enters into sub-contracts and supply contracts with the nominated sub-contractors and suppliers, respectively, he/she has no right to determine the amounts payable to them. As such, the nominated sub-contractors' and nominated suppliers' payment statement should be submitted to the quantity surveyor for assessment through the contractor within seven days before the submission of payment application by the contractor under the Contract. The contractor is required to pay the nominated sub-contractors and nominated suppliers within 14 days the amounts specified for them in each interim certificate for which the contractor receives payment from the employer (clause 29.7(2)). Figure 4.1 illustrates the payment time frame established under the Contract.

4.4 Payment mechanism

Several parties are involved in the interim valuation process, and their contractual obligations are interrelated. For example, the quantity surveyor's interim valuation supports the architect's preparation of interim certificates, and the architect is bound to issue interim certificates to

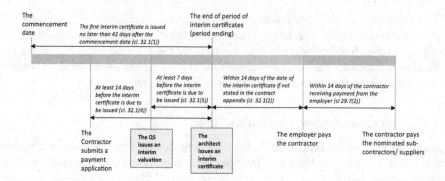

Day: Calendar days include Sundays and general holidays under the General Holidays Ordinance (Chapter 149, Laws of Hong Kong)

Figure 4.1 Time frame for preparing an interim certificate

the employer to make payments to the contractor. As such, the Contract delineates different requirements to the employer, architect, quantity surveyor, contractor, nominated sub-contractors and nominated suppliers. The contractual processes establish a systematic payment mechanism that facilitates efficient interim payments to ease the contractor's cash flow.

4.4.1 Procedures and requirements for a payment valuation

The quantity surveyor is not the only party to undertake interim payment valuation. In this regard, it is critically important for the quantity surveyor to discuss the valuation arrangement with the contractor and other consultants at the project outset so that the required obligations can be met within the time frame stipulated in the Contract. For instance, the quantity surveyor and the contractor should reach agreement on the schedule and format of the payment applications.

The quantity surveyor is also required to inform the building services engineer of the interim payment timeline and agree with him on the deadline for his valuation of building services installations if the terms of the quantity surveyor's appointment exclude such valuation duties. Moreover, the quantity surveyor should submit his valuation to the architect at least seven days before the interim certificate is due to be issued.

If the payment term specified under the Contract is monthly progress payments, the contractor is obliged to submit the monthly payment application with a statement containing sufficient information for the quantity surveyor's assessment. The statement is largely based on measured items in the BQ, and it reflects the payment amount of the completed work and the materials/goods delivered to the site to be claimed by the contractor up to the period end of an interim certificate. The statement includes not only the estimated gross amount of the work in progress but also a priced list of the materials/goods delivered to or adjacent to the site or stored off-site according to the case (clause 32.1(4)).

The payment applications from the nominated sub-contractors and nominated suppliers are also included in the statement. To allow efficient interim valuation, the contractor is required to provide all of the accounts, vouchers, receipts, delivery notes and other documents that may reasonably be required by the quantity surveyor for assessment (clause 32.1(4)). After the date of substantial completion, the contractor should submit the payment application as and when further payments are required.

   The basic purpose of the interim valuation is to estimate the 'gross' valuation of the work in progress and the materials/goods delivered to the site, i.e., the accumulated value of the work completed, not the value of the work completed in that period. To determine the estimated gross value, the quantity surveyor usually applies a percentage assessment against each BQ item based on the site inspection records. However, this approach is not applicable to the payment of the preliminaries and the on-site materials/goods because of their distinctive expenditure patterns. For the preliminaries, the amount of each item contained in the bill of preliminaries is apportioned under three headings: 1) initial cost, 2) running cost and 3) removal cost. The running cost is further split into time-related and work-related costs. Each apportioned amount is then assessed and included in the interim valuation. For the payment of the on-site materials/goods, the quantity surveyor should estimate the quantities of the materials/goods delivered to or adjacent to the site and value them at the appropriate supply rates based on the unit rates priced by the contractor in the BQ.

   Table 4.2 lists the build-up of the gross valuation and summarises the items to be added and deducted from the gross valuation as stipulated under clauses 32.2 (3) and (4).

Table 4.2 Items to be included in and deducted from the gross valuation

Items To Be Included in the Gross Valuation under Clause 32.2 (3)	*Items To Be Deducted from the Gross Valuation under Clause 32.2 (4)*
a) Permanent work properly carried out including variations ordered	a) (i) Correction of errors in setting out the works under clause 7(2)
b) Temporary work properly carried out where their value is included as a separate sum in the BQ	(ii) Replacement or reconstruction of materials, goods, or work under clause 8.3(c)
c) Preliminary items properly provided or carried out where their value is included as a separate sum in the BQ	(iii) Rectifying defects under clause 17.3(5)
d) Materials or goods on-site	
e) Materials or goods off-site under clause 32.3	b) Decreases in the costs of labour and/or materials under clause 38 (if applicable)
	c) Any other deduction amounts required by the Contract
f) Work carried out by nominated sub-contractors	
g) Contractor's tender accepted for work provided for by a prime cost sum under clause 29.4	
h) Materials or goods supplied by nominated suppliers	
i) (i) Profit on nominated sub-contractors and nominated suppliers	
(ii) Attendance upon nominated sub-contractors	
j) (i) Statutory fees and charges under clause 6.3	
(ii) Opening up and testing materials, goods, etc. under clause 8.2	
(iii) Insurances resulting from the employer's failure to insure under clause 22B.2(2) or 22C.3(2)	
k) Direct loss and/or expense under clause 27.2	
l) Increases in the costs of labour and/or materials under clause 38 (if applicable)	
m) Any other amounts required by the Contract	

To determine the amount due to the contractor in an interim certificate, the retention and the total amounts previously certified are subtracted from the gross valuation (clause 32.2(1)). Based on this concept, a formula for the interim valuation is as follows:

The amount due in an interim certificate = Gross valuation – Retention – Total amounts previously certified

In addition, the employer is entitled under the Contract to make any deduction from the amount due to the contractor under an interim certificate provided that a notice is served to the contractor stating the details of the deduction (clause 32.1(6)).

An example of a hypothetical build-up of payment valuations from the first interim certificate to the final certificate based on the above formula is shown in Table 4.3. It is assumed that the contractor carries out the superstructure works of a small-scale residential project with the awarded contract sum HK\$1,000 million. The contract period and defects liability period are 550 days and 12 months respectively. The term of payment is a monthly progress payment and the retention rules follow the provisions stated in the Contract. The limit of retention is 5% of the contract sum, excluding nominated sub-contract sums and nominated supply contract sums, plus the retention held in respect of the nominated sub-contractors and nominated suppliers. The works have been substantially completed by the contractor, and the architect issues the Substantial Completion Certificate at the end of the contract period. The architect also issues the Defects Rectification Certificate upon the expiry of the defect's liability period. For simplicity, it is assumed that the final contract sum is the same as the awarded contract sum.

After receiving the payment application from the contractor, the quantity surveyor should conduct a preliminary check on the application to ensure that it includes no errors or unreasonable items. The quantity surveyor can review the extent of the payment claimed and plan for the site visit. Site visits are a key activity in an interim valuation because they enable the quantity surveyor to validate the contractor's payment applications.

During a site visit, the contractor's representative (usually the contractor's quantity surveyor) will accompany the quantity surveyor to examine the completed work and the unfixed materials and goods delivered to the site with reference to the submitted application. The representatives of the nominated sub-contractors and nominated suppliers should attend the site visit if they have submitted their payment applications to the contractor. The quantity surveyor, as the assessor, should check the reasonableness of the contractor's payment valuation claim, such as the percentage of the reinforced concrete frames completed and the quantities of reinforcing bars delivered to site. Any queries on the scope of the completed work done and the unfixed materials and goods delivered should be resolved with the contractor's quantity surveyor and the representatives of the nominated sub-contractors and nominated suppliers during the site visit where possible. It is advisable for the quantity surveyor to take photos during the site visit for future reference.

After the site visit, the quantity surveyor should prepare the interim valuation based on the contractor's payment application, which includes a cost breakdown statement abstracted from the BQ such as preliminaries, measured work and unfixed materials/goods. The quantity surveyor assesses the amount of each item based on the site records and other substantiations

Table 4.3 Example of a hypothetical build-up of payment valuations

	Interim Certificate 1	Interim Certificate 2	... Certificate 10	Interim Certificate 11	Interim Certificate 12	... Certificate 17	Interim Certificate 18	... Final Certificate
Estimated gross value of work done + on-site materials/goods (MC)	10,000,000	16,000,000	380,000,000	480,000,000	580,000,000	792,000,000	800,000,000	800,000,000
Estimated gross value of work done + on-site materials/goods (NSC/NS)	2,000,000	3,000,000	95,000,000	120,000,000	145,000,000	198,000,000	200,000,000	200,000,000
Gross Valuation ^	**12,000,000**	**19,000,000**	**475,000,000**	**600,000,000**	**725,000,000**	**990,000,000**	**1,000,000,000**	**1,000,000,000**
Retention (MC)	1,000,000	1,600,000	38,000,000	40,000,000#	40,000,000#	40,000,000#	20,000,000*	0**
Retention (NSC/NS)	200,000	300,000	9,500,000	10,000,000#	10,000,000#	10,000,000#	5,000,000*	0**
Less: Retention	**1,200,000**	**1,900,000**	**47,500,000**	**50,000,000**	**50,000,000**	**50,000,000**	**25,000,000**	**0**
	10,800,000	17,100,000	427,500,000	550,000,000	675,000,000	940,000,000	975,000,000	1,000,000,000
Less: Previously Certified	0	10,800,000	315,000,000	427,500,000	550,000,000	925,000,000	940,000,000	975,000,000
Amount Due	10,800,000	6,300,000	112,500,000	122,500,000	125,000,000	15,000,000	35,000,000	25,000,000

MC: Superstructure main contract NSC: Nominated sub-contract NS: Nominated supply contract
^ Gross valuation includes the other items stated in clause 32.2(3) and 32.2(4)
# Limit of the retention is reached
* Half of the retention is payable to the contractor upon the issue of the Substantial Completion Certificate
** Remaining half of the retention is payable to the contractor upon the issue of the Defects Rectification Certificate

offered by the contractor and marks up his/her comments directly on the contractor's submitted statement.

It is normal practice for the quantity surveyor to inform the contractor's quantity surveyor of the assessment results before issuing the formal valuation to the architect. Any disagreement with the application submitted by the contractor's quantity surveyor should be discussed. The intention of such discussion is not to negotiate an agreement with the contractor on the payment amount but to avoid any mistake arising from the quantity surveyor's misinterpretation of the contractor's submitted statement. Finally, the quantity surveyor concludes the valuation and sends his/her recommendation to the architect together with a statement of sums payable to the nominated sub-contractors and nominated suppliers.

4.4.2 Payment for nominated sub-contractors and nominated suppliers

The quantity surveyor is required to assess the payments claimed by the nominated sub-contractors and nominated suppliers. Under the Contract, the interim payments to them are made through the contractor. As such, the quantity surveyor is obliged to prepare a statement to the contractor that shows the amount of the interim payment included for each nominated sub-contractor and nominated supplier and the amount retained by the employer. The quantity surveyor should also send individual payment notifications to the nominated sub-contractors and nominated suppliers (clause 29.7(1)). The notifications should include the payment details so that the nominated sub-contractors and nominated suppliers can serve written notice to the contractor if they have not received payment within the time frame stipulated in the Contract.

Under the Contract, the architect has the discretion to require reasonable proof from the contractor that the payments to the nominated sub-contractors and nominated suppliers included in previous certificates have been made (clause 29.8(1)). It is suggested that reasonable proof would be a payment receipt or letter from the nominated sub-contractor or nominated supplier concerned. If the contractor withholds any payment due, cannot substantiate the reasons for such withholding and has not issued a notification to the nominated sub-contractor or nominated supplier concerned, the architect is obligated to certify that the contractor has failed to comply with the request to furnish the substantiation (clause 29.8(3)). Upon receipt of the certification, the employer can pay the nominated sub-contractor or nominated supplier concerned directly and deduct the same amount from any sum due to or to become due to the contractor (clause 29.8(3)).

As stated above, the Contract sets out a special mechanism for direct payment to the nominated sub-contractors and nominated suppliers rather than through the contractor. This contract provision indicates that the nominated sub-contractors and nominated suppliers are in a very different position than domestic sub-contractors and domestic suppliers and that the architect is not involved in their payment rights under normal circumstances. Figure 4.2 shows a typical payment flow under the Contract. The figure illustrates the payment flow from the upstream to the downstream and shows how the nominated sub-contractors and nominated suppliers are protected under the Contract if their payment is unreasonably withheld by the contractor.

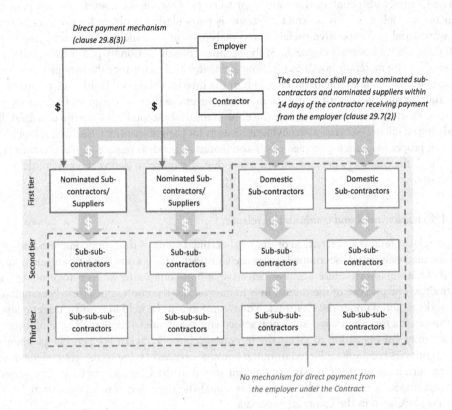

Figure 4.2 A typical payment flow from the employer to the contractor and his sub-contractors

4.4.3 Payment for materials/goods on-site and off-site

The gross valuation includes the estimated value of materials/goods on or adjacent to the site in accordance with the contract provision of clause 32.2(3). Delivery to the contractor's yard is not qualified for payment valuation unless that yard is adjacent to the site. This provision can ease the contractor's cash flow and encourage the contractor's early (but not premature) supply of materials so that late deliveries and unexpected shortages are less likely to occur and hold up progress.

The Contract specifies three restrictions on situations in which the materials/goods are to be paid by the employer in interim certificates. First, they must be intended for inclusion in the permanent work under the Contract. This excludes formwork and construction plant or other consumable stores. Second, the materials/goods must not have been prematurely brought to or adjacent to the site. The term 'prematurely' can be interpreted based on the contractor's master programme approved by the architect. Third, the materials/goods must be adequately protected against weather, other damage and theft. It should be noted that once these materials/goods are certified and paid for, they become the employer's property, although the contractor remains responsible for any loss or damage to them.

The Contract also enables the contractor to be paid for off-site materials/goods prior to their on-site delivery. This contract provision is particularly important for projects involving substantial and expensive prefabricated materials off-site, such as structural steelwork materials. Under Contract clause 32.3, the architect has discretionary power to include the value of off-site materials/goods in interim certificates and to instruct the quantity surveyor to estimate the value of the materials/goods in the interim valuation. Unlike the payment of materials on site, the quantity surveyor must take several actions to safeguard the employer's interest. First, it is essential to request the contractor to substantiate ownership of materials/goods stored off-site, such as with payment receipts for the sale contract. Second, reasonable proof of proper insurance coverage of off-site materials/goods is required from the contractor. Third, the quantity surveyor should make due allowance to defray the cost of delivery to the site.

4.4.4 Retention rules and conditions of release

The employer retains a percentage of the total certified value of the work completed on-site and the materials/goods ascertained under the Contract. This is known as a retention, a safeguard of the employer's interest against defective or non-conforming work or materials by the contractor. The purpose of the retention is to ensure that the contractor properly completes the works required under the Contract. The retention is to be held in trust by the employer for the contractor, nominated sub-contractors and nominated suppliers. The employer is exempt from any obligation to invest the retention (clause 32.5(1)). For periodic payments, the retention is regularly deducted from the amount certified as due to the contractor in an interim certificate until the maximum amount stated in the Contract appendix is reached. The percentage of certified value retained is called the retention percentage (often 10%), which is also stated in the Contract appendix.

The retention calculation for the nominated sub-contractors and nominated suppliers may not be the same as that of the contractor because of different terms of payment in the nominated sub-contracts or nominated supply contracts. For example, the stage payments of a nominated sub-contract for lift installation may give rise to irregular retention deduction. As such, the retention of payment for the nominated sub-contractor's work in progress and the nominated supplier's supply of materials/goods should be calculated in accordance with the nominated sub-contract and the nominated supply contract, respectively, and then added to the retention held on the payment for the contractor's work (clause 32.4(3)). According to clause 32.4(1), additional payment for direct loss/or expense under clause 27 and cost adjustment for fluctuations of labour or materials under clause 38 are not subject to retention.

The retention on the payment for the contractor's work in progress is calculated by applying the retention percentage to the estimated gross valuation (clause 32.4(1)). The retention held by the employer should not exceed the limit stated in the Contract appendix (clause 32.4(2)). Because retention is calculated differently for the main contract, nominated sub-contracts and nominated supply contracts, as mentioned above, the retention limit is a summation of the respective maximum retentions of the contractor, the nominated sub-contractors and nominated suppliers. The retention limit is usually 5% of the contract or sub-contract sum, unless another percentage is agreed upon and entered into the Contract appendix.

The retention held by the employer will be released according to the contract provisions. The Contract sets out the conditions required for the payment of the retention.

The retention is released to the contractor in two stages. Half of the retention is payable to the contractor upon the issue of the Substantial Completion Certificate. The payment certificate for this first moiety of the retention should be issued within 14 days of Substantial Completion of the whole of contract works, or a section or a relevant part (clause 32.5(2)). Afterwards, the amount of any interim certificate issued after Substantial Completion will still be subject to withholding of the remaining half of the retention until the Defects Rectification Certificate is issued. The payment of the second moiety of the retention should be made under an interim certificate within 14 days of the issue of the Defects Rectification Certificate for the whole of contract works, or a section or a relevant part (clause 32.5(4)). This abatement of the half of the retention represents that the contractor's liability under the Contract has been discharged because the remaining half retention held by the employer is for the interest of the employer because of the risk and defects in the defects liability period.

With regard to a section or a relevant part, the amount of the retention shall be deemed to bear the same relationship to the retention held for the whole of contract works, as the estimated amount contained in the contract sum for that section or relevant part bears the contract sum (clause 32.5 (3)). Figure 4.3 illustrates the payment time frame established in the Contract.

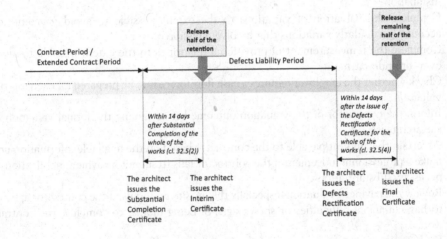

Figure 4.3 Time frame for the release of the retention

4.4.5 Interim valuation practice

Interim valuation is an estimate of the amount due in an interim certificate. Although the interim certificate is still not final and is subject to review in the final certificate, there is no further interim payment if the contractor becomes insolvent and its employment is terminated. Therefore, every interim payment must be a reasonable amount for both the employer and the contractor. The term 'reasonable' means that the payment amount is not only fair for the contractor but also safeguards the employer's interests. A high valuation will neither benefit the employer nor the project as a whole. If the valuation is low, it will place unreasonable financial burden on the contractor, as its cash flow will be affected.

However, the quantity surveyor must be mindful of any special payment request from the contractor. For example, it is a Hong Kong custom for the contractor to ask for the interim payment in advance or for a higher payment amount in the interim certificate before the Chinese New Year. If payment advancement or higher valuation is given, it may create a risk to the employer, who may pay sums with no benefit, and the employer may incur a financial loss if the contractor becomes insolvent. Thus, approval from the employer is essential before the arrangement of valuation of any advanced payment.

To prepare a fair and efficient interim valuation, the quantity surveyor should conduct the following practice in the interim valuation.

1 As necessary, apply the adjustment percentage arising from the arithmetical error or commercial discount in the tender submitted by the contractor when preparing the interim valuation.
2 Agree to a breakdown of preliminaries with the contractor for the assessment of the initial costs, running costs and/or removal costs for each item of the preliminaries.
3 Consult with the clerk of works regarding questionable materials delivered on-site and/or hidden work to ensure proper valuation of the interim payment.
4 Inform the building services engineer of the time frame of the interim valuation if he/she participates in or is responsible for preparing the interim valuation for building services installation.
5 Complete the valuation of variations to the extent possible to avoid payments on account, particularly variations that involve omission of work.
6 Complete the remeasurement of provisional quantities to the extent possible to avoid over- or underpayment.
7 Check whether the employer requires a cash flow forecast to be prepared for each interim valuation.
8 Inform the contractor of the valuation outcome before issuing the formal valuation to the architect.
9 Ascertain the amount payable to the contractor based on the available information and make a professional judgement if the contractor fails to submit a payment application or necessary substantiations.
10 Review the payment valuation, especially the on-site materials, if the contractor appears to have financial difficulties or shows signs of being unable to complete the contract works.
11 Evaluate the remaining amount of the anticipated final contract sum after each payment valuation during the final construction stage to ensure that the contractor is not overpaid.

4.5 Chapter summary

Interim valuation is an estimate of the amount due to the contractor. The architect is contractually bound to issue interim certificates at the period stated within the contract period or any extended contract period or at agreed-upon intervals after the completion of contract works. The quantity surveyor is responsible for providing a reasonable and fair valuation so that the employer can make timely and adequate interim payments to the contractor. The architect and the quantity surveyor are obligated to issue interim certificates and valuations respectively, and the contractor is responsible for providing sufficient information for the valuation. An interim valuation is not expected to be completely accurate, and a balance must be struck between protecting the employer's interests and ensuring that the contractor is not starved of cash flow. However, any interim valuation should be as accurate as possible to avoid under- or over-certification. Many factors affect the accuracy of interim valuation, such as the extent of the quantity surveyor's experience and level of valuation skill, the correctness of the contractor's payment application, the available information for assessing variations ordered and other contractual claims and the remeasurement progress of provisional quantities. Thus, accurate valuation is even more important to provide the contractor with sufficient payment to complete his work and to ensure that the employer has not overpaid the contractor, as overpayment is unlikely to be recovered in some bankruptcy cases.

4.6 Case study

Case one: the employer's remedy to the contractor's failure to obtain a surety bond

A contractor fails to deliver the surety bond in the month after the tender award. What is the employer's remedy under the Contract? Is the employer eligible to deduct the bond amount from the contractor's payment?

Discussion

It is the contractor's responsibility to obtain a surety bond before the start of the construction works. The contractor should approach an insurance company or a bank approved by the architect and arrange the bond execution according to the specimen contained in the contract documents. The bond value is stated in the Contract appendix and is usually 10% of the contract sum. However, if the contractor fails to deliver the bond within 28 days of the tender acceptance, the employer may withhold an amount not greater than the bond value as a cash security (clause 33.3). The quantity surveyor should take this into account when preparing the interim payment valuation. According to the Contract, the amount withheld should be released if a bond is submitted by the contractor. The quantity surveyor should incorporate the release of the cash security into the interim valuation following the delivery of the bond.

Case two: assessment of the interim payment for the preliminaries

The contractual arrangement of a residential project is a lump sum contract with the BQ, and progress payment is adopted as the term of payment. As the quantity surveyor for this project, you are checking the first application for interim payment submitted by the contractor's quantity surveyor. You note that he/she has applied a single percentage to all of the BQ items, including the preliminaries, in the payment application. How do you assess the preliminaries in the interim valuation?

Discussion

The bill of preliminaries contains various items, and their expenditure patterns can be very different. Instead of applying a single percentage against each item of the preliminaries in the interim valuation, the quantity surveyor usually apportions the amount of each item under three headings: initial cost, running cost and removal cost. The running cost is further split into time-related and work-related costs. Although there is no definite rule for such apportionment, the quantity surveyor and the contractor's quantity surveyor should agree on the apportioned amounts at the project outset. Once a consensus is reached, the apportionment will be adopted in each interim valuation. Figure 4.4 shows how each item is apportioned based on its nature and monthly expenditure in the payment assessment.

The examples for the above cost splitting are given to explain the rationale for apportioning the costs of the preliminaries in the interim valuation. Site management primarily includes the salaries of the contractor's site management staff, who are paid regularly throughout the contract period. As such, the cost is related to the time factor only, and the payment is made on the proportion of the days lapsed over the overall contract period.

The extent of the plant, scaffolding and tools provided by the contractor during the contract period is significantly related to contract works carried out on-site. This work-related item is paid based on the value of the completed contract works as a proportion of the contract sum excluding preliminaries and PC sums.

The contractor is also required to establish site accommodations, such as a site office and meeting room, and to maintain them during the contract period. An initial cost is required for the provision of the site accommodation, and a time-related cost is required to sustain daily operations, such as for consumables, cleaning and maintenance. After completion of

BQ Abstract			Payment Assessment			
Bill No. 1 – Preliminaries			Initial Cost	Running Cost		Removal Cost
BQ Item	Description	Amount		Time-related	Work-related	
		HK$	HK$	HK$	HK$	HK$
1/3/B	Site management	11,000,000		11,000,000		
1/4/C	Plant, tools, scaffolding, etc.	30,000,000			30,000,000	
1/4/G	Labour	14,000,000		14,000,000		
1/5/D	Site accommodation	5,000,000	3,000,000	1,000,000		1,000,000
1/5/F	Temporary water	600,000	200,000		400,000	
1/7/A	Temporary electricity	5,000,000	1,000,000		4,000,000	
1/9/D	Removal of water	400,000			400,000	
1/11/F	Cleaning	5,000,000			2,000,000	3,000,000
1/11/G	Telephone and broadband service	400,000		400,000		
1/13/C	Insurance	30,000,000	20,000,000	10,000,000		
1/15/A	Surety bond	2,000,000	1,800,000	200,000		
1/15/D	PCF and CITA levies	2,400,000			2,400,000	
1/15/F	Setting-out	2,000,000			2,000,000	
1/16/D	Hoarding	14,400,000	9,000,000		4,400,000	1,000,000
	Total:	122,200,000	35,000,000	36,600,000	45,600,000	5,000,000

Figure 4.4 Example of the apportionment of individual items in the bill of preliminaries

the project, the contractor should pay for the site accommodation to be dismantled and removed. Both the initial and removal costs are paid to the contractor according to the actual progress of establishing and dismantling the site accommodation.

Furthermore, the contractor is required to substantiate the premium receipts for the payment of insurance and surety bonds. The usual practice is to certify the exact amount of the premium paid by the contractor at the beginning of the project. If any of the costs of the preliminaries are left after the contractor is paid, they will be paid for over the contract period. Alternatively, the whole BQ amount of insurance and surety bonds can be spread over the contract period for payment notwithstanding the entire premium paid by the contractor. However, this may negatively affect the contractor's cash flow, particularly as the BQ amounts of insurance and surety bonds share a large portion of preliminaries.

Case three: imposition of liquidated and ascertained damages (LAD)

The contractor's progress of contract works is affected by various factors such as inclement weather, late instruction from the architect and poor site management. After reviewing the notices of claims submitted by the contractor, the architect grants the contractor a 15-day extension according to the Contract. The contractor then completes the works on 31 July 2020, the architect certifies this date as the Date of Substantial Completion. The completion date stipulated in the Contract is 3 July 2020 and the liquidated and ascertained damages (LAD) rate is HK$50,000 per day. Advise the employer about the imposition of LAD on the contractor's payment.

Discussion

The contractor failed to complete the works by the completion date stipulated under the Contract and is liable to pay or allow the employer to deduct the LAD at the stipulated rate for the period between the completion date/extended completion date and the date of substantial completion. According to the Contract, the employer has the power to recover damages from the amount due to the contractor, provided that a non-completion certificate has been issued by the architect (clause 24.2) and a notice has been served to the contractor at least seven days before the deduction is made (clause 40). LAD is a genuine pre-estimate of the expected loss or damage to the employer as a result of a delay in completion. The quantity surveyor should remind the employer of his right to impose LAD and indicate the deductible amount in the interim valuation. The following example shows how the LAD amount is calculated.

Extended Completion Date:

3 July 2020 (Completion Date) + 15 days (EOT) = 18 July 2020

Delay:

31 July 2020 (Date of Substantial Completion) − 18 July 2020 = 13 days

Amount of LAD:

13 days x HK$50,000/day = HK$650,000

Case four: valuation of on-site materials

The contractor's quantity surveyor has submitted a payment application for the work completed and the materials delivered to the site. You note that the contractor's quantity surveyor has also submitted delivery notes and invoices to support the payment claim for on-site materials. As a quantity surveyor, advise your assistant on how to assess payment for on-site materials.

Discussion

The Contract allows payment to be made to the contractor for the materials delivered to the site but not yet fixed into position. In deciding what on-site materials to be certified for payment, the quantity surveyor must examine the payment application submitted by the contractor and ask the contractor's quantity surveyor to provide necessary substantiations. During site inspection, the quantity surveyor can seek advice from the clerk of works regarding the materials' compliance with the contract requirements and specifications. The payment for on-site materials depends on the nature of the materials, but the materials delivered to the site should be properly stored and protected in any case. The quantity surveyor should check the supply quantities of the complying materials according to the specifications, the supplier's delivery notes and other available information, such as the product details shown on the packaging. Any materials not intended for inclusion in the contract works (e.g., formwork) or that is prematurely brought to the site should not be paid. The quantity surveyor should refer to the contractor's approved master programme to decide whether the materials have been delivered prematurely. Payment for on-site materials can be finalised by applying the appropriate wastage factors and supply rates with reference to the BQ rates for the quantities after on-site verification.

All work should be measured as fixed in its place in the BQ according to the HKSMM4 (i.e., the principle of 'measured net'), so the all-in rates in the BQ are deemed to include all necessary cutting and waste. However, the supply quantities of the on-site materials include the contractor's wastage, which should be deducted from the interim valuation. Wastage is usually represented as a percentage factor that applies to certain types of work. Depending on the type and nature of the work, different wastage factors for various materials should be adopted in the interim valuation.

The payment amount for on-site materials can be determined by pricing the supply quantities after deducting the wastage. The contractor's substantiations can support the pricing of on-site materials. Suppliers' invoices and sales contracts are examples of valid substantiations that indicate the supply rates. However, the quantity surveyor cannot simply rely on the contractor's information without making judgements during the assessment. For example, the given supply rates should be reasonable when compared with the supply rates proportioned from the all-in rates in the BQ. Overpayment for on-site materials may incur a financial loss to the employer if the contractor becomes insolvent after receiving the money, particularly if the remaining BQ amounts are insufficient for the new contractor to fix the materials into the work. Hence, if the quantity surveyor finds any irregularities in the supply rates during the assessment, it is practicable to countercheck them by obtaining quotations from suppliers in the market. Finally, the accumulated value of on-site materials should be deducted from the gross valuation when those materials have been fixed into the work by the contractor because the completed work is paid instead.

5 Contractual claims

Learning outcomes

Upon completion of this chapter, you should be able to do the following.

1 Understand the general cost reimbursement provisions under the Standard Form of Contract.
2 Understand the general principles for assessment of variations.
3 Understand the general rules of (i) prolongation and disruption and (ii) direct loss/or expense claims.
4 Understand the procedures for (i) submission of notice of delay and (ii) claims for additional payment associated with time extensions.
5 Know the general claim headings for the submission of construction claims related to time disruption.

5.1 Introduction

This chapter provides a general understanding of cost reimbursement under variation assessment and contractual claims related to time disruption. Relevant construction contract books that provide further details are suggested. This chapter discusses the clauses in the 2005 edition of the Agreement & Schedule of Condition of Building Contract for use in the Hong Kong Special Administrative Region.

A claim is simply an assertion of a party's right under the terms and conditions of contract or under law. The majority of construction works in Hong Kong are carried out with the use of standard forms of contract such as:

1 Agreement & Schedule of Conditions of Building Contract for use in the Hong Kong Special Administrative Region 2005 Version (with quantities) – for private sector use generally ("the Contract" hereinafter).
2 The Government of the Hong Kong Special Administrative Region – General Conditions of Contract for Building Works, 1999 Edition – for government of the HKSAR use generally.

DOI: 10.1201/9781003212355-5

5.2 Types of contractual reimbursement claims under the private sector's standard building contract

Three main contractual reimbursement claims under the Contract, which are:

5.2.1 *Assessment of variations ordered during the course of the contract*

Variations ordered during the course of the Contract include the following:

1   alteration to the type, standard or quality of any of materials/goods to be used in the contract works;
2   addition, substitution or omission of any project works;
3   alteration of the kind or standard of any of materials/goods to be used; and
4   alteration or modification of the

   (i)   design,
   (ii) quality or
   (iii) quantity of the contract works.

'Contract works' are briefly described in the Articles of Agreement and are shown upon, described by or referred to in the Contract; they include any change made to the works in accordance with the Contract.

   Under the Contract, variation works are defined as follows.

Clause 1 Interpretation and definitions

Variation: a change instructed by the Architect to the design, quality or quantity of the Works including:

   (i)   *an alteration to the type, standard or quality of any of the materials or goods comprising the Works;*
   (ii) *the addition, substitution or omission of work; and*
   (iii) *the removal from the Site of materials or goods and the demolition and removal of work except where provided for in the Contract or where the materials, goods or work are not in accordance with the clause 8.1; or the imposition of an obligation or restriction instructed by the Architect regarding:*
   (iv) *access to the Site or use of any parts of the Site;*
   (v)   *limitation of working space:*
   (vi) *limitation of working hours; or*
   (vii) *the sequence of carry out or completing works*

or the addition or alteration to or omission of such obligations or restrictions imposed by the Contract.

The rules for the valuation of variation under the Contract are stipulated under clause 13.4 as follows.

Clause 13.4 Valuation rules

(1) *Where the Valuation relates to the carry out of:*

    (a) *additional or substituted work which can be properly valued by measurement;*

    (b) *work which is the subject of Provisional Quantities or Provisional Items; or*

    (c) *work involved in the expenditure of Provisional Sums,*

Where the variation relates to the carrying out of the above clauses 13.4 (1) (a) to (c), the work should be measured and valued in accordance with the rules stipulated in clauses 13.4(1) (i) to (iv), which are tabulated below:

Valuation related to the use of BQ rates

Table 5.1 Valuation related to the use of BQ rates or fair rates

Contract Clause	Variation Work Ordered	Valuation Method	Remarks
13.4(1)(c)(i)	a) Is the same as or similar in character. b) Is carried out under the same or similar conditions to work priced in the BQ. c) It does not substantially change the quantity of that work.	The rates in the Contract bill (BQ rates) shall determine the valuation.	Example 1
13.4(1)(c)(ii)	a) It is the same as or similar in character to work priced in the BQ. b) It is not carried out under the same or similar conditions. c) It substantially changes the quantity of that work.	a) The rates in the Contract bill (BQ rates) shall determine the valuation, with b) fair adjustment for the difference in conditions or quantity.	Example 2
13.4(1)(c)(iii)	a) It is not the same as or similar in character to any work priced in the BQ.	Valued at <u>fair rates</u>.	Example 3
13.4(1)(c)(iv)	The word 'conditions' in clause 13.4(1) means physical conditions and not financial conditions.	Reminder only	

Variation related to the use of daywork rates

The rules and guidelines for the valuation of variation using daywork rates are summarised in Table 5.2.

Table 5.2 Valuation related to the use of daywork rates

Contract Clause	Variation Work Ordered	Valuation Method	Contractor Follow-up Action	Remarks
13.4(2)	Valuation cannot be properly measured and valued under clause 13.4(1).	a) Work may be carried out as daywork (with prior consent of the architect). b) Work is valued at prime cost, including labour, materials, goods, plant and equipment plus overhead and profit	*Clause 13.4(2)(a)* Gives at least 7 days' notice to the architect before work is carried out or, where work is urgently required, as much prior notice as is practicable.	Example 4
			Clause 13.4(2)(b) Submits vouchers (specifying (i) time spent, (ii) workers' names and (iii) materials, goods, plant and equipment) to the architect for verification within 14 days of the work being carried out.	
			Clause 13.4(2)(c) Price at the daywork rates in the BQ. *Clause 13.4(2)(d)* If there are no daywork rates in the BQ, the following pricing is used. i) Labour rates: the average daily wages of workers engaged in government building and construction projects published by the Census and Statistics Department. ii) The net cost of materials and goods plus the cost of packing carriage and delivery. iii) The hiring cost of plant and equipment together with the cost of transportation, fuel, maintenance and insurance. iv) (1) Additional percentages for overhead and profit (based on the prime cost of the labour, material, goods and plant and equipment in the BQ) or (2) Additional 15% if the above percentage is not allowed.	

Variation related to omission

The rules and guidelines for the valuation of variations related to omissions are summarised in Table 5.3.

Table 5.3 Variations related to omission

Contract Clause	Variation Work Ordered	Valuation Methods	Remarks
13.4(3)(a)	Work included in the BQ.	The rates in the Contract bills (BQ rates) shall determine the valuation.	Example 5
13.4(3)(b)	Omission of work for which, in the quantity, surveyor's opinion, the contractor incurred reasonable expense.	A fair adjustment shall be made to the valuation with respect to the expense.	Example 6

Variation related to other matters

Valuation of variations related to other matters are summarised in Table 5.4.

Table 5.4 Variation related to other matters

Contract Clause	Variation Work Ordered	Valuation Methods

13.4(4)	a) Work not related to
	  (i) additional or substituted work, or
	  (ii) omission of work.
	b) Work related to other matters not involving measured work, such as the imposition of or change to an obligation or restriction.
	c) The rules in clause 13.4(1), (2) or (3) cannot be applied.
		Fair valuation shall be made. Under clause 13.4(5), appropriate allowance shall be made in valuation under this section, e.g., a percentage or lump sum adjustment made to the Contract bills.

13.4(6)	a) Work compliance with variation instructed (clause 13.1). b) A deemed variation substantially changes (clause 14.3) the conditions. c) Work carried out results in BQ rates becoming unreasonable or inapplicable.	Use new rates based on the BQ rates with adjustment for fair allowance for the difference in the conditions.
13.4(7)	a) Work in addition to daywork valuation (clause 13.4(2)). b) Work that cannot be measured and valued (clause 13.4(1)).	With the contractor's agreement, the architect may instruct the contractor to carry out the work and value it on a daywork basis (clause 13.4(2)).
13.4(8)	a) Variation for additional work after substantial completion.	Fair valuation shall be made.
13.4(9)	a) Work for which the contractor can be reimbursed by payment under any other provision of the Contract.	No allowance is made for (i) direct loss and/or expense due to delay to the progress of the works, (ii) disruption; or (iii) any other cause.

Examples of valuation of variation are summarised in Table 5.5 (illustrate Examples 1,2,3,5 and 6)

Table 5.5 Examples of valuation of variation (*Examples 1, 2, 3, 5 and 6*):

EXAMPLE BACKGROUND INFORMATION

Information abstracted from BQ

SUSPENDED CEILINGS: INTERNAL
An 'ABC' acoustic tile false ceiling or approved equivalent;
  600 x 600 x 16 mm thick complying with BS476; Part 4:
  including an XYZ 15 mm grid system and all necessary
  fixing accessories, painting etc. as per the specification.

	Suspended ceilings with a grid system				
A	Supply only	200	m²	560	112,000
B	Allow fixing, wastage and profit, etc.	100	m²	110	11,000
C	Allow fixing, wastage and profit, etc. 3.5–5.0 m above floor	100	m²	180	18,000

VALUATION OF VARIATION

Example	Variation	Valuation of Variation	Cost Amount HK$
1	Use BQ rate Add 30 m² acoustic tile to Classroom no. 1.	Ceiling tile: BQ rate (1/A) 30 m² x 560 = HK$16,800	HK$16,800 HK$ 3,300
		Fixing: BQ rate (1/B) 30 m² x 110 = HK$3,300	HK$20,100
2	Use BQ rate with fair adjustment Add 20 m² acoustic tile to Classroom no. 2, 6 m above the floor.	Ceiling tile: BQ rate (1/A) 20 m² x 560 = HK$11,200 Fixing: BQ rate (1/B) with fair adjustment n.e. 3.5 m is HK$110 3.5–5.0 m is HK$180 a) Difference between the above two items: HK$180 – HK$110 = HK$70 Assume 15% for more difficult work above 5 m in height (the % is not fixed; it depends on the level of job complexity)	HK$11,200 HK$ 5,750 HK$16,950

Example	Variation	Valuation of Variation	Cost Amount HK$
		b) <u>New fixing rate is</u> difference between the two stages: $(180 + 70) + (180 + 70)$ X (15%) = HK\$287.5 Fixing: Pro-rata rate as above 20 m² X HK\$ 278.5 = HK 5,750	
3	<u>Use the new rate</u> Supply and fix the 'bright' metal ceiling to Classroom no. 3. This new item is assumed to be different from or similar in character to any work priced in the BQ.	<u>Methods that can be adopted</u> 1. Use a similar item from a previous project as a rate reference with fair adjustment. 2. Get a quotation from the supplier. Assuming that the quotation from the supplier and the checked receipt from the contractor are used, it costs HK\$1,000 per m² (including supply and fixing) <u>The new rate is</u> HK\$1,000 + 15% (profit and overhead for the Contractor) = HK\$1,150	HK\$1,150
5	<u>Omission Work</u> Omit 10 m² 'ABC' acoustic tile false ceiling in Classroom no. 4.	Direct omission work Ceiling tile: BQ rate (1/A) 10 m² x 560 = HK\$5,600 Fixing: BQ rate (1/B) 10 m² x 110 = HK\$1,100	<u>Omission Variation</u> HK\$5,600 HK\$1,100 HK\$6,700
6	Omit 25 m² 'ABC' acoustic tile false ceiling in Classroom no. 5. a) The instruction is too late. The 10 m² ceiling tile material has been purchased and delivered to site and is ready for installation. b) Ceiling tile: BQ rate (1/A) 10 m² x 560 = HK\$5,600; c) No need to deduct.	<u>Omission works</u> a) In the QS's opinion, the contractor incurred reasonable expense. b) This instruction entailed a loss to the contractor due to purchased material (the incurred expense on the ceiling tile, as it is assumed that none of these materials can be used). c) Maintain the material (ceiling tile unchanged): Ceiling tile: BQ rate (1/A) not yet purchased 15 m² × 560 = HK\$8,400 (needs to be omitted) d) Omission of fixing work Fixing: BQ rate (1/B) 25 m² × 110 = HK\$2,750	HK\$8,400 HK\$2,750 HK\$11,150 Net valuation (Omission) of the architect's instruction

The valuation method by using daywork is summarised in Table 5.6 (illustrates Example 4).

Table 5.6: Example 4: Valuation by daywork

Background information
During the door installation process, the door opening D1 (900 mm x 2,100 mm) was constructed. The architect issued an Architect Instruction to revise D1 to D2 (1,500 mm x 2,100 mm) as the opening was being constructed; thus, widening work on door opening is required. It is assumed that the BQ has the daywork bill, which was priced separately. The operation involves the following steps.

1	Break up the existing plastered concrete wall (150 mm thick) to widen the door opening – use the daywork rate.
2	Revise the opening to the required size – use the daywork rate
3	Perform the necessary protection work for the adjacent work completed – material cost for protection screen, e.g., PVC sheet + the daywork rate.
4	Make good every disturbed work, e.g., concrete work, plastering work, etc. – use the daywork rate.
5	If the door is not yet fixed, omit door D1 and add door D2 (the rate for D2 is in the BQ). As no demolition work is needed for D1, use the BQ rate to omit D1 and add D2.

Valuation of the variation work (by daywork)

Step	Daywork	Material	
1	Allow 3 man days (general labour) HK450/day (rate from BQ) x 3 = HK$1,350.		HK$1,350
2	Allow 1 man day (general labour) HK450/day (rate from BQ) x 1 = HK$450 and 1 man day (plasterer) HK650/day (rate from BQ) x 1 = HK$650.		HK$450 HK$650
3	Allow 1 man day (general labour) HK450/day (rate from BQ) x 1 = HK$450.	Allow: PVC sheet as protection screen 10m^2 x HK$60/m^2 (material cost = HK$600)	HK$450 HK$600
4	Allow 2 man days (plasterer) HK650/day (rate from BQ) x 2 = HK$1,300.		HK$1,300
	Add 15% profit and overhead for items 1 to 4 ($4,800 x 0.15).		HK$720
5	Omission: D1 @$2,200 (BQ rate) Addition: D2 @$4,500 (BQ rate)		(HK$2,200) HK$4,500
	Total cost for this variation		HK7,820

5.2.2 Claims for extension of time and related prolongation claim

Time-related claims

Variation is not the only category of claims for the contractor; there are also claims for time extensions. The granting of time extensions for delayed events prevents the deduction of liquidated damages from the contractor provided that such events are included in the Contract provisions (clause 25) as listed events. In addition, the contractor can submit a claim if he/she has incurred or is likely to incur a direct loss or expense under qualified events, as stipulated in clause 27.

1   Events that qualify for time extensions

Clause 25.1(3) of the contract specifies the following events for which the contractor is entitled to claim an extension of time (EOT).

Contract Clauses	Listed Qualified Events Entitled for EOT
25.1(3)(a)	force majeure;
25.1(3)(b)	inclement weather conditions, being rainfall in excess of twenty millimetres in a 24-hour period (midnight to midnight) as recorded by the Hong Kong Observatory station nearest to the Site, and/or its consequences adversely affecting the progress of the Works;
25.1(3)(c)	the hoisting of tropical cyclone warning signal No. 8 or above or the announcement of a Black Rainstorm Warning;
25.1(3)(d)	the Excepted Risks;
25.1(3)(e)	loss and/or damage caused by a Specified Peril;
25.1(3)(f)	an Architect's instruction under Cl 2.4 (ambiguities in documents) to resolve an ambiguity, discrepancy in or divergence between the documents listed in that clause;
25.1(3)(g)	an Architect's instruction under Cl 8.2 (inspection and testing) requiring the opening up for inspection of work covered up or the testing of materials, goods or work and the consequential making good where the cost of that opening up, testing and making good is required by that clause to be added to the Contract Sum;
25.1(3)(h)	an Architect's instruction under Cl 13.1 (A.I. for variation) requiring a Variation;
25.1(3)(i)	an Architect's instruction under Cl 13.2 (A.I. for Provisional Quantities; Provisional Items and Provisional Sum) resulting in an increase in the work to be carried out of sufficient magnitude to cause delay, provided that the variance was not apparent from the Contract Drawings;
25.1(3)(j)	an Architect's instruction under Cl 23.3 (postponement or suspension) regarding: (i) the postponement of the Date for Possession of the Site or part of the Site; (ii) the postponement of Commencement Date of the whole or part of the Works; or (iii) the postponement or suspension of the whole or a part of the Works, unless: • notice of the postponement or suspension is given in the Contract; or • the postponement or suspension was caused by a breach of contract or other default by the Contractor or any person or whom the Contractor is responsible;

(Continued)

(Continued)

Contract Clauses	Listed Qualified Events Entitled for EOT
25.1(3)(k)	compliance with Cl 34.1 (effect of finding antiquities) or with an Architect's instruction under Cl 34.2 (Architect's instruction concerning a fossil, antiquity or defect) requiring the Contractor to permit the examination, excavation or removal by a third party of an object of antiquity found on Site;
25.1(3)(l)	late instructions from the Architect, including those to expend a Prime Cost Sum or a Provisional Sum, or the late issue of the drawing details, descriptive schedules or other similar documents referred to in Cl 5.6 (further drawings, details, descriptive schedules and similar documents) except, to the extent that the Contractor failed to comply with Cl 5.7(2) (documents to be provided to Contractor on time);
25.1(3)(m)	delay caused by a delay on the part of Nominated Sub-contractor or Nominated Supplier in respect of an event for which the Nominated Sub-contractor or Nominated Supplier is entitled to an extension of time under the sub-contract or supply contract;
25.1(3)(n)	delay caused by a sub-contractor or supplier nominated by the Architect under Cl 29.2(6) despite the Contractor's valid objection subject to Cl 29.2(7);
25.1(3)(o)	delay caused by the nomination of a replacement Nominated Sub-contractor or Nominated Supplier under Cl 29.13 (re-nomination) including any prolongation of the period of the relevant subcontractor or the time for the supply and delivery of materials and goods, provided that the determination of the employment of the original Nominated Sub-contractor or the termination of the original Nominated Supply Contract was not in the opinion of the Architect a consequence of a breach of contract or other default by the Contractor or any person for whom the Contractor is responsible;
25.1(3)(p)	delay caused by a Specialist Contractor;
25.1(3)(q)	delay caused by a statutory undertaker or utility company referred to in CL.6.4(1) (statutory undertakers and utility companies) failing to commence or to carry out its work in due time provided that the Contractor has taken all practical measures to cease it to commence and to carry out and complete its work on time;
25.1(3)(r)	the failure of the Employer to supply or supply on time materials, goods, plant or equipment that he agreed to provide for the Works;
25.1(3)(s)	the failure of the Employer to give possession of the Site or, under Cl 23.1(2), (commencement and completion) a part of the Site on the Date for Possession of the Site or the part of the Site stated in the Appendix, or the Employer subsequently depriving the Contractor of the whole or a part of the Site;
25.1(3)(t)	delay to the Works due to time not reasonably foreseen by the Contractor in obtaining approval or consent from a Government department;
25.1(3)(u)	a special circumstance considered by the Architect as sufficient grounds to fairly entitle the Contractor to an extension of time; and
25.1(3)(v)	an act of prevention, a breach of contract or other default by the Employer or any person for whom the Employer is responsible.

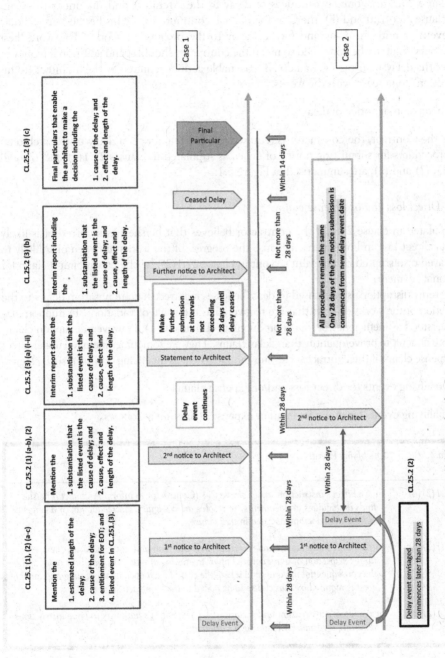

Figure 5.1 Submission of notice of delay under the Contract

2   Contractor's actions to submit Extension of Time claims

The contractor has duties to take the following actions (i) under the Contract clause 25.1, and clause 25.2 to submit two notices of delay to the Architect and (ii) under the Contract clause 25 (4)(a) and (b), the Contractor shall continuously use his/her best endeavours to prevent or mitigate delay and further delay to the progress of Works. The word 'best endeavours' shall not be construed to mean the contractor should spend additional money to recover the delay. Contractor shall do all reasonable actions required by the architect (to his satisfaction) to proceed with the works.

3   Submission of notice of delay

Under the Contract, the contractor is required to submit a notice of delay to the architect. The procedures for submitting notice of delay, as stipulated in clauses 25.1(1) and (2) and 25.2(1), (2) and (3), are summarised in Figure 5.1.

5.2.3 Direct loss and/or expense claims

As stipulated in clause 27.1, if the contractor believes that he/she has incurred or is likely to incur direct loss and/or expense because the progress of the works has been or is likely to be delayed or disrupted by an event set out in clause 27.1(2), he/she can submit a claim for additional payment.

The term 'disruption' is often used to describe losses incurred that are not from delays to the completion of the works or costs that are recovered as the value of variations. Disruption may be understood as inefficiency or loss of productivity (David, 2011). However, disruption claims are more difficult to prove/quantify than delay claims. They are sometimes included under direct loss/expense claims if the contractor can substantiate them as qualifying events in such claims.

1   Qualifying events for direct loss and/or expense claims

The qualifying events for direct loss and/or expense claims are as follows.

Contract Clause	Qualifying Events
Cl 27.1(2)(a)	an Architect's instruction under clause 2.4 (Contractor to inform Architect if he/she finds ambiguities in documents) to resolve an ambiguity, discrepancy in or divergence between the documents listed in that clause;
Cl 27.1(2)(b)	an Architect's instruction under clause 8.2 (inspection and tests) requiring the opening up for inspection of work covered up or the testing of materials, goods or work and the consequential making good where the cost of such opening up, testing and making good is required by that clause to be added to the Contract Sum;
Cl 27.1(2)(c)	an Architect's instruction under clause 13.1 (Architect's authority to issue instructions requiring a variation) requiring a Variation;

Contract Clause	Qualifying Events
Cl 27.1(2)(d)	an Architect's instruction under clause 13.2 (instruction for provisional quantities, provisional items and provisional sums) resulting in an increase in the work to be carried out of sufficient magnitude to cause delay or disruption, provided that the variance was not apparent from the Contract Drawings;
Cl 27.1(2)(e)	an Architect's instruction under clause 23.3 (postponement or suspension) regarding: (i) The postponement of the Date for Possession of the Site or a part of the Site; (ii) The postponement of the Commencement Date of the whole or a part of the Works; or (iii) The postponement or suspension of the whole or a part of the Works, unless: • notice of the postponement or suspension is given in the Contract; or • the postponement or suspension was caused by a breach of contract or other default by the Contractor or any person for whom the Contractor is responsible;
Cl 27.1(2)(f)	Compliance with clause 34.1 or with an Architect's instruction under clause 34.2 requiring the contractor to permit the examination, excavation or removal by a third party of an object of antiquity found on the site;
Cl 27.1(2)(g)	late instructions from the Architect, including those to expend a Prime Cost Sum or a Provisional Sum, or the late issue of the drawing details, descriptive schedules or other similar documents referred to in clause 5.6, except to the extent that the Contractor fails to comply with clause 5.7(2);
Cl 27.1(2)(h)	delay or disruption caused by a sub-contractor or supplier nominated by the Architect under clause 29.2(6) despite the Contractor's valid objection, subject to clause 29.2(7);
Cl 27.1(2)(i)	delay or disruption caused by a Specialist Contractor;
Cl 27.1(2)(j)	the failure of the Employer to supply or supply on-time materials, goods, plant or equipment that he agreed to provide for the Works;
Cl 27.1(2)(k)	the failure of the Employer to give possession of the Site or, under clause 23.1(2) (commencement and completion), a part of the Site on the Date for Possession of the Site or the part of the Site stated in the Appendix, or subsequently the Employer depriving the Contractor of the whole or part of the Site; and
Cl 27.1(2)(l)	any other delay or disruption for which the Employer is responsible including an act of prevention or a breach of contract.

2 Submission of notice of claims for additional payment

As stipulated in clause 27.1(1), the contractor can claim additional payment on direct loss and/or expense if the progress of the works has been or is likely to be delayed or disrupted by a qualifying event listed in clause 27.1(2). To do so, he/she must follow the procedures set out in clause 28, as detailed in Figure 5.2.

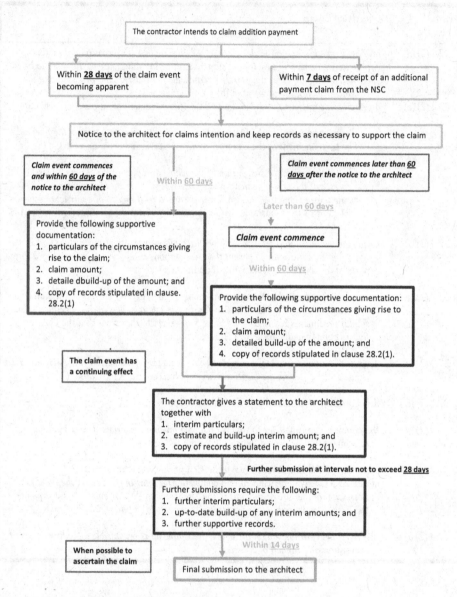

Figure 5.2 Notices of claims for additional payment

5.3 General claim headings

The Contract stipulates the provisions for the contractor to claim extension of time and the loss and/or expense due to disruption under clause 25 and 27. The procedures of a claim's submission are also well elaborated hereunder.

5.3.1 Additional payment

Additional payment in regard to the direct loss and/or expense shall include the following steps as stated below:

1   additional expense that can be easily identified and substantiated (e.g., material and labour);
2   increased site overhead/preliminaries due to prolongation of the Contract period;
3   non-recoverable fluctuations (e.g., cost escalation);
4   head office overhead and profit;
5   finance charges;
6   cost claims from nominated sub-contractors;
7   depreciation of assets; and
8   delayed release of retention.

5.3.2 Costs for which additional payment claims cannot be made

Additional payment claims cannot be made for the following expenses:

1   cost of accelerating the works;
2   cost of overtime; and
3   cost of preparing a claim.

5.4 Evaluation of direct loss and/or expense

5.4.1 Additional expense

Additional expenses are normally individual costs that have not been valued under the valuation of variation. However, such expenses should be checked separately to ensure that they do not overlap with other expenses or costs claimed under separate valuations (e.g., extra plant hire time that might be covered under preliminaries for a prolonged contract period). If such an expense is assessed based on raw cost, 15% overhead and profit is normally acceptable.

Table 5.7 Example 7: Assessment of additional expenses

Background information
The architect issues an instruction to the contractor to suspend all on-site work for one day and to assist in arranging a ceremonial visit from the employer's management. The architect therefore grants a one-day EOT. The contractor quotes all the actual expenses from the ceremonial visit (which is within the extended period granted under the EOT). The quoted actual expenses of HK$50,000 are considered to be reasonable and acceptable to all consultants.
Assessment of the additional expenses is as follows: HK$50,000 (actual expenses) + HK$7,500 (15% overheads and profit) = HK$57,500

5.4.2 Increased site overhead/preliminaries

Increased site overheads/preliminaries, what we call a "Prolongation Claim", is where the claim is based on extra site overhead costs due to prolongation of the Contract period. This evaluation is made on the extended construction period. The common method used to reimburse a contractor is to adopt the preliminaries breakdown provided by the contractor for the time related and non-time related items. The one-off expenses (e.g., establishment of site offices) will not be considered.

Table 5.8 Example 8

Preliminaries Breakdown

BQ Ref: (1)	Description (2)	Total Cost HK$ (3)	Initial Cost HK$ (4)	Time Related Cost HK$ (5)	Work Related Cost HK$ (6)	Completion Cost HK$ (7)
1/22/B	Temporary roads	48,000	20,000	20,000	-	8,000
1/24/D	Contract's superintendence and staff	2,000,000	-	2,000,000	-	-
1/24/F	Setting-out	700,000	350,000	350,000		-
1/25/B	Care of the works	330,000	-	330,000	-	-
1/26/H	Compliance with enactments and regulations	100,000	-	80,000	20,000	-
1/29/C	Clearance of the Site on completion	100,000	-	-	-	100,000
1/44/C	Site accommodation for the Contractor	1,500,000	500,000	900,000	-	100,000
1/45/C	Office for the Supervising Officer	1,300,000	400,000	800,000	-	100,000
1/46/A	Hoardings and gantries	350,000	50,000	290,000	-	10,000
1/46/B	Signboards	40,000	10,000	25,000	-	5,000
1/46/D	Lighting and power	700,000	200,000	450,000	-	50,000
1/51/A	Tests	500,000	200,000	-	250,000	50,000
	Total	7,668,000	1,730,000	5,245,000	270,000	423,000

In this example, the preliminaries bill with cost implications are abstracted and divided into seven main items:

(1)   BQ reference – the reference of the abstracted item;
(2)   item description – the scope of the item;
(3)   total cost – the total cost of the selected item;
(4)   initial cost – the set-up cost of the selected item, which is valued and included in the first payment valuation after the Contract commences;
(5)   time-related cost – the cost of the item related to the time, which is valued and included in the monthly payment during the contract period;
(6)   work-related cost – the payment valuation will only cover the cost once this item of work or portion of the work is completed; and
(7)   completion cost – the payment valuation will be valued when the work is completed, and the work is demolished and/or removed from the site.

Assessment of Prolongation Cost Based on the Preliminaries Bill:

(1)   Payment for preliminaries
(2)   In the first payment
(3)   During the period of interim payment
(4)   The last payment

The assessment of the direct loss/expense for the time-related items will be valued in proportion to the EOT granted.

Background: (1) Construction period is 365 days; and (2) Granted EOT: 10 days

Prolongation cost assessment:

HK$5,245,000 (total time-related cost) divided by 365 days x 10 days of EOT
= HK$143,698.63

5.4.3 Non-recoverable fluctuations

With a fixed price contract that has no provision for the fluctuations in the costs of labour and materials, there is no mechanism for adjusting the Contract sum for those costs. It is deemed the fluctuation costs have been allowed by the Contractor in his tendered rates and/ or cost for all necessary costs for the proper execution of work during the original Contract period. Thus, the costs of escalation attributable to the prolonged contract period (outside the original contract period) should be reimbursed to the Contractor. Therefore, the claim under this situation is entitled.

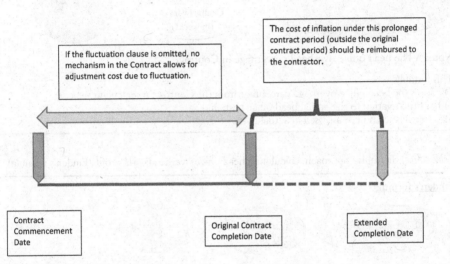

If the fluctuation clause is omitted, no mechanism in the Contract allows for adjustment cost due to fluctuation.

The cost of inflation under this prolonged contract period (outside the original contract period) should be reimbursed to the contractor.

Contract Commencement Date

Original Contract Completion Date

Extended Completion Date

Figure 5.3 Non-recoverable fluctuations

5.4.4 Head office overhead and profit

The profits of the head office are not only gained from one particular construction project but also from other projects in the same company that generate profits. Under the same principle, the head office overhead also contributes to every project in the same company.

Components of head office overhead

The components of the head office overhead often include the following:

1   purchase or rent and rates for offices, plant and yards;
2   maintenance and operating costs;
3   director's salaries and expenses;
4   head office technical staff salaries and 'on costs' (e.g., surveyors, planners, purchasers and the like);
5   head office administrative staff salaries and 'on costs' (e.g., accountants, typists, messengers, cleaners, maintenance staff and the like);
6   administrative expenses (e.g., postage, printing, stationery, telephones and the like);
7   travelling expenses, including the provision of motor cars for directors and other staff; and
8   legal and professional fees.

Assessment

There are different methods to assess the head office overheads and profits. In Hong Kong the formulae approach is usually adopted by some employers:

Table 5.9(i) Formulae approach: Calculating head office overheads and profit (Hudson's formula)

Hudson's formula

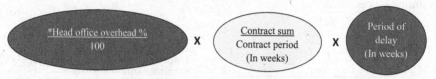

Note: * The head office overhead percentage in Contract

This formula
1) adopts the head office overhead percentage from the Contract to calculate the cost;
2) has little relation to the actual head office cost; and
3) is unrealistic and not assessed in actual costs.

Table 5.9(ii) Formulae approach: calculating head office overheads and profit (Emden's formula)

Emden's formula

Note: *The head office overhead = Total overhead cost + profit/Total turnover x 100%

This formula
1) is an attempt to improve upon the Hudson's formula
2) uses the actual overhead/Profit % rather than the one contained in the Contract

Table 5.9(iii) Formulae approach: calculating head office overheads and profit (Eichleay formula)

Eichleay formula (based on the actual loss)

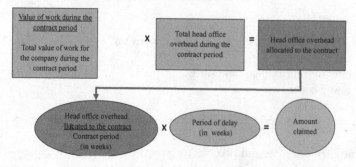

Note:
1) This formula is calculated by comparing the value of work carried out for the project in the contract period with the value of all of the work carried out by the contractor as a whole in the contract period.
2) A share of head office overhead for the contractor can then be allocated using the same ratio and expressed as a lump sum for the particular contract.

Other indirect additional notes

Loss and/or expense under this heading and those not sufficiently "direct" attributes to the project usually are not acceptable without specific proof and agreement. No assessment/valuation shall be made if it is an indirect cost (e.g., possible loss of new contracts).

5.4.5 Finance charges

Additional financing charges incurred due to an extended contract period are recoverable as direct loss and expense. Interest rates should be calculated at actual and real rates.

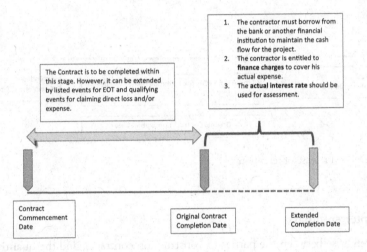

Figure 5.4 Finance charges

5.4.6 Cost claims from sub-contractors

The contractor may incur liabilities to the nominated sub-contracts or supply contracts as a result of an event for which he/she is entitled to recover direct loss and/or expense. This would be a legitimate direct cost to be reimbursed to the nominated sub-contractors or suppliers under the Contract. However, the onus of proof and substantiation is on the contractor and the nominated sub-contractors and suppliers.

5.4.7 Depreciation of assets

Where a costs claim arises from a substantial delay to the Contract period, the contractor's plant and equipment will be in use longer than planned. The contractor thus incurs depreciation in the value of those assets; this loss would be recoverable.

5.4.8 Delayed release of retention

Delay to the extension of time granted under the Contract could result in a delay in the release of the retention. Thus, if there are proven incurred finance charges on the retention money held by the Employer, they are a direct expense arising from the disruption to the completion of the Works and the Contractor is entitled to recover the actual expense. It is the same principle as that of the recovery of finance charges as in section 5.4.6.

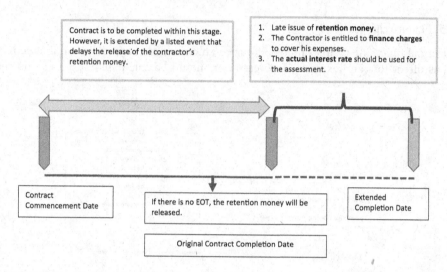

Figure 5.5 Delayed release of retention

5.5 Chapter summary

Claims often arise between the parties to construction contracts, and the quantity surveyor assists with the valuation of these claims. This chapter highlighted the following salient points that the quantity surveyor should keep in mind when assessing contractors' claims on construction projects.

1   Prolongation costs arise as a consequence of delay.
2   Disruption costs arise as a consequence of adverse effects on productivity.
3   Works may be disrupted without delaying the Contract completion date.

Table 5.10 Example 9

- Instruction issued to omit certain works and add other works.
- No delay to the overall completion date.
- Require the contractor to re-schedule the works sequence and revise overall additional expenses incurred by the site layout.
- In this case, the Contractor is entitled to recover his/her cost from the disruption.

4   Effects of disruption/loss of productivity

To illustrate the effects of disruption/loss of productivity, the contractor must substantiate the following:

(i)   that the contractor's original programme and sequence of works were seriously affected by the employer;

(ii) that the contractor was prevented from carrying out the work according to the original programme; and

(iii) the works were disrupted without any delay of the contract completion date.

5   Key factors in successful disruption claims include the following.

(i)   The contractor must submit all necessary contemporary records, including

a)   daily disruption schedules and
b)   the site diary.

(ii) They must be verified and signed by site supervising staff (e.g., clerks of works, resident engineers, etc.) to establish an accurate record.

6   Failed disruption claims

Disruption claims may fail for the following reasons.

(i)   The contractor did not keep all necessary contemporary records as substantiation.

(ii) There is no formal record verified by the site supervising staff from the employer.

5.6 Case study

For a case study, please refer to Examples 1 to 9 in this chapter.

Notes to reader:

1   Examples are for reference only

---

2   Contract clauses are abstracted from the Agreement & Schedule of Condition of Building Contract for use in the Hong Kong Special Administrative Region 2005 Edition published by the Hong Kong Institute of Architects, The Hong Kong Institute of Construction Managers, and The Hong Kong Institute of Surveyors.

6 Cost control and monitoring

Learning outcomes

Upon completion of this chapter, you should be able to do the following.

1 Understand the uses of cost control and monitoring techniques.
2 Know the procedures for adopting cost control and monitoring techniques.
3 Know how to prepare a financial report (example is provided).
4 Know how to prepare a final account (example is provided).

6.1 Introduction

As the custodian of the employer's budget, the quantity surveyor should ensure that the authorised project cost budget is not exceeded. Construction costs are a significant portion of a project's costs. To keep construction costs under control, the quantity surveyor must use cost control and monitoring techniques in the building design and construction processes. This chapter addresses cost control and monitoring in connection with construction projects and then provides a case study.

6.1.1 Cost control

Cost control refers to techniques used to control expenditures on building projects so that they remain within a pre-determined budget(s) from the design to the construction stages. Cost control commences in the Inception Stage and ends when the project is handed over to the employer.

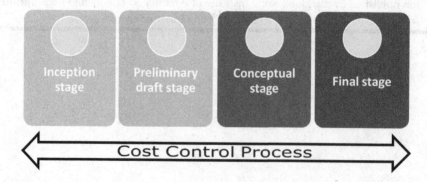

Figure 6.1 Cost control process

DOI: 10.1201/9781003212355-6

6.1.2 *The purpose of cost control*

The goal of cost control is to limit the employer's expenditure to a pre-set amount. It also seeks to ensure that the final contract sum is within the budget. Cost control may involve a balance of design expenditures on various elements of the project. It enables the employer to have a monetary value for the project. Its major components are summarised in Table 6.1.

Table 6.1 The components of cost control

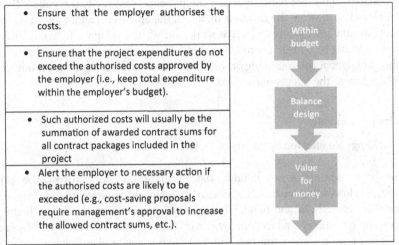

• Ensure that the employer authorises the costs.
• Ensure that the project expenditures do not exceed the authorised costs approved by the employer (i.e., keep total expenditure within the employer's budget).
• Such authorized costs will usually be the summation of awarded contract sums for all contract packages included in the project
• Alert the employer to necessary action if the authorised costs are likely to be exceeded (e.g., cost-saving proposals require management's approval to increase the allowed contract sums, etc.).

6.2 Cost planning and control strategy

A cost control strategy typically consists of three stages:

Stage 1 – setting the budget;
Stage 2 – cost planning and control of design development; and
Stage 3 – controlling the expenses at the procurement and construction stages.

6.3 Stage 1 – setting the budget

The employer's needs, objectives and targets must be considered before the budget is set. In general, there are four types of employers.

1   An occupier (e.g., a factory owner) wants a building that meets his/her particular manufacturing needs. Profit is obtained indirectly from the benefits derived from the manufacture of product inside the building, and the open market value of the building is not a concern.

2   A property company (e.g., a company constructing a building to sell) wants to make a direct financial profit from the development. The accuracy of the budget greatly affects this profit.

3   An investor (e.g., a company constructing a building to rent) wants to make a direct financial profit from rental income from the building. The accuracy of the budget greatly affects this profit.

4   A local authority (e.g., the hospital authority building a hospital for the community) must be publicly accountable for the money spent. Profit cannot be valued and can only be assessed by a cost-benefit analysis. The project should maximise the value for money.

6.4 Stage 2 – designing to meet the budget

After fixing the budget, the project team (e.g., the architect, the structural engineer, the building services engineer, etc.) designs the project, considering the following areas.

6.4.1 Design method

In some cases, the employer does not impose a preliminary budget limit on the design team, and the designers have free hand to design the project. (This method is called 'Budgeting a Design'.) However, if, when the design is finalised, the employer cannot afford the construction cost, this may lead to unproductive work and wasted designs. To avoid such a situation, the 'Designing to Budget' approach should be used. Under this approach, the employer establishes a budget at the beginning of the design stage, and the design team must prepare a design according to the budget authorised.

6.4.2 Design economics

Cost of buildings are affected by variety of factors.

1 Plan shape – the shape of a building has an important effect on cost: the simpler the shape, the lower the construction costs. Irregular and complex shapes may also affect construction methods and buildability, increasing construction costs. A higher wall to floor ratio (perimeter wall to floor area) will also increase the quantities of some perimeter items (e.g., concrete, finishes, windows). An irregular plan shape may also result in increased costs for setting out site works and drainage works.
2 Building size – a larger building will have reduced elemental unit costs (i.e., the construction cost per square foot).
3 Circulation space (e.g., entrance halls, passages, corridors, stairways, lift wells, etc.). This 'dead space' cannot be used for profitable purposes while incurring heating, cooling and lighting costs.
4 Storey height – when storey height increases, the costs of all vertical elements (e.g., concrete walls, wall finishes, vertical building services pipe, external envelope, etc.) is also increased.

6.5 Stage 3 – controlling costs

6.5.1 Cost control/checking during drawing production stage

The detailed cost control and checking processes in the pre-contract stage are illustrated in Section 6.6.

6.5.2 Cost control in the procurement stage

Commonly used procurement methods include (i) cost reimbursement procurement, (ii) price-in-advance procurement and (iii) management-based contracts. Price-in-advance procurement can most effectively provide price certainty and control of variation.

Table 6.2 Types of procurement methods

A) Cost Reimbursement Procurement

   1 Cost plus Percentage/Fixed Fee

      | Original construction Cost | + | Profit (Cost) |

   2 Target Cost

      | Contractor to build within the target cost |

B) Price in Advance Procurement
   1 Lump Sum Contract
      1.1 Contract based on measured BQ
      1.2 Drawing and Specifications

      | Pricing the construction cost from drawings plus specifications |

   2 Re-measurement contract
      2.1 Bills of approximate quantities

   3 Design and Build

      | Contractor takes a design role Construct role |

C) Management-based Contract
   1 Management Contracting
   2 Construction Management

      | Include the contractor in the consultant team during the design stage |

6.5.3 Cost/time control in the construction stage

Effective cost/time control in the post-contract stage includes joint efforts from the project team members and the quantity surveyor, client and contractor. Success can be achieved through the control of (i) variation, (ii) claims, (iii) progress of work, (iv) proper recording of all information and (v) timely and accurate reporting on the project's financial status (e.g., financial statements and draft final accounts) by the quantity surveyor to the employer. Details on preparing these documents are illustrated in Examples 2 and 3.

6.6 Cost control process in connection with a construction project

The cost control processes are summarised in Figure 6.2, and detailed explanations are then provided. Cost control includes all methods of controlling the cost of construction projects within the limits of the budget, from the inception of the design scheme to final completion of the contract works. The control process can further be divided into pre-contract and post-contract stages.

6.6.1 Pre-contract cost control

The pre-contract cost control process involves the following steps.

1   Preliminary estimate

    (i)   This is the first indicator of the overall construction cost of the proposed project.

    (ii) It must be a reasonable estimate: it cannot be too low because the design is at a preliminary stage and design development is still in progress, and it cannot be too high because it may exceed the employer's budget.

    (iii) No drawings are available, only preliminary information.

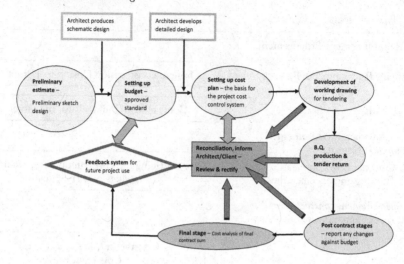

Figure 6.2 Summary of the cost control process

2   Preliminary cost plan

   (i)   This includes an evaluation of the designer's first sketch designs.
   (ii) It includes an evaluation of alternative designs, so the employer can choose the
         most economical or optimal solution.
   (iii) It includes a preliminary cost budget.

3   Detailed cost plan, which forms the basis of the cost control system.

   (i)   The final sketch design is completed.
   (ii) The client reviews and approves the design.
   (iii) A cost plan is produced based on the final design.
   (iv) Sufficient details and description are included.

4   BQ production

   BQ is measured when the tender drawings are developed. The quantity surveyor uses
      the measured BQ and prices it based on current market price data to provide a
      pre-tender estimate that will serve as a reference for comparison during the tender
      analysis stage.

5   Cost checking

   Cost checking exercises are very important and must be carried out for every esti-
      mate/cost plan prepared or updated. The goal of cost checking is to determine
      whether the budget has been exceeded. If it has, reconciliation exercises must
      be conducted, with justifications and suggestions referred to the client/design
      team for a further decision. This process does not seek to stifle or dictate the

choice of design but aims to avoid dead-end works. The final cost checking process is to be conducted when all working drawings have been completed, as they are the basis of the BQ. The pre-tender estimate can be used to evaluate the returned tenders. Any abnormal pricing in the tendered items or trades should be highlighted, and the client and design team should be informed via the tender report.

[Remarks: Estimation methods are detailed in Chapter 1]

6.6.2 Post-contract cost control

The cost control processes are to be carried out routinely throughout the construction stage and the final account stage after the construction contract has been awarded and the construction works have commenced.

Construction stage

1   The construction stage involves the following cost control processes.

(i) Cost control of variations issued: The quantity surveyor must conduct a cost check on any variation and inform the architect and employer of the costs of any changes in their design. They must advise whether the change in design is reasonably valued.

(ii) Valuation of payment: The quantity surveyor is required to prepare interim valuations (i.e., interim payments) for the works completed. Completed works on-site and materials on- or off-site must be valued with high accuracy. Any payment submission from the contractor should be assessed as soon as possible and be included in the interim payment.

2   The financial statement should inform the architect/employer of the financial status of the following items:

(i)   main contract/sub-contract/direct contract;
(ii) cost of the variations;
(iii) submitted claims from the contractor; and
(iv) anticipated final contract sum.

3   The financial statement has the following purposes.

(i)   To keep the client regularly informed (usually monthly or quarterly) of their financial commitments to the Contract (i.e., whether the cost is greater than or less than the contract sum).
(ii) To reconcile major recent changes (i.e., to compare the latest financial statement with previous statements).
(iii) To report the current situation and forecast future cost status (i.e., final expenditure).

(iv) To assist in the monthly valuation of works in progress using a detailed summary of the estimated values of the Architect's instructions (i.e., assessment of variations).

(v) To serve as an integral part of the final account.

4   Guidelines for preparing a financial statement are as follows.

(i)   All figures included should be as accurate as possible.

(ii) The quantity surveyor should seek agreement on the costs of the variations with the contractor and report the agreed-upon sums in the monthly or quarterly financial statement.

(iii) Estimated cost effects of building services works provided by the building services engineer should be shown separately on the financial statement.

(iv) If there are possible losses and/or expenses for prolongation or disruption claims, due allowance must be made under a separate heading.

Two examples are illustrated in this section. Example 1 (Table 6.3) is the normal summary page of a financial statement/report for the use by the practitioner. Example 2 (Table 6.4) is in exercise format: the reader can complete the financial report based on the information provided. The answer to this exercise is also provided in Table 6.5.

Table 6.3 Example 1: Financial statement in report format

Objective: To give a sample of a financial statement prepared by a QS

Proposed Residential Development
Main Contract
Financial Statement as at 1 Dec 2020

	OMISSIONS HK$	ADDITIONS HK$
Awarded Contract Sum		108,000,000.00
Omit Contingencies	5,000,000.00	
Adjustment of Provision Sums	3,000,000.00	2,500,000.00
Adjustment of Prime Cost Sums	22,000,000.00	20,000,000.00
Adjustment of Provisional Quantities	2,000,000.00	3,000,000.00
Adjustment of Prime Cost Rates	1,500,000.00	1,300,000.00
Adjustment of Issued Variations	500,000.00	3,200,000.00
Adjustment of Anticipated Variations	-	1,000,000.00
	34,000,000.00	139,000,000.00
Less OMISSIONS		(34,000,000.00)
Anticipated Final Contract Sum		105,000,000.00
		==========
Remaining Contingencies		**3,000,000.00**

Table 6.4 Example 2: Preparation of financial report (exercise)

Project Background Information

1) *Project name*: Robinson International School at Sai Kung, Hong Kong
2) *Client*: Richman Academic Association
3) *Contract procurement form*: Lump sum with Bills of Quantities
4) *Main contractor*: WellBuild Construction Co. Ltd. (WB)
5) *Budget for the whole project*: HK$91,000,000.00, including the working packages in item 6 below
6) *Contract works packages*:
   6.1 Foundation contract: Final contract sum settled at HK$10,000,000.00
   6.2 Direct fitting out contract: Contract sum at HK$1,000,000.00
   6.3 Main contract: Contract sum at HK$80,000,000 (including HK$3,000,000 in contingencies for future variation use)
7) *Information abstracted from bills of provisional and prime cost sums*

	Description	Qty	Unit	Rate	$	¢
	Bill No. 10					
	Provisional and Prime Cost Sums					
	Provisional Sum					
	The following provisional sums may be expended in whole or in part as directed by the architects or wholly deducted from the contract sum if not required:					
A	Provide the sum of $250,000.00 for construction of a fence wall along northern site boundary				250,000.	00
B	Provide the sum of $600,000.00 for signage				600,000.	00
	Prime cost sums for nominated sub-contractors					
	Provide the following prime cost sums for the supply and fixing of works by specialist contractors who shall be nominated by the architect and who are declared to be nominated sub-contractors to the main contractor in accordance with clause 27 of the Conditions of Contract (the main contractor shall refer to clauses xxx of the Specification – Preliminaries for attendance as described).					
C	Provide the sum of $4,000,000.00 for Lifts Installations				4,000,000.	00
D	Add __1__ % for profit				40,000.	00

(Continued)

Table 6.4 (Continued)

E	Add for attendance as described	40,000.	00
F	Provide the sum of <u>$7,000,000.00</u> for Electrical and HVAC Installation	7,000,000.	00
G	Add __1__ % for profit	70,000.	00
H	Add for attendance as described	70,000.	00
I	Provide the sum of <u>$6,000,000.00</u> for Fire Services Installation	6,000,000.	00
J	Add __1__ % for profit	60,000.	00
K	Add for attendance as described	60,000.	00

8) **Variations, claims summary and assessment status**

A.I. No.	Description	ACC's estimate (HK$)		Contractor's claimed amount (HK$)	Remarks
		Omission	Addition		
	Architectural Instruction				
MC1	Commencement of Contract	(No Cost Effect)		–	
MC2	Revised architectural drawings at 1/F	(163,000.00)	193,300.00	60,000.00	Under negotiation
MC3	Revised aluminum windows (W2)	(1,980.00)	2,300.00	450.00	Agreed
MC4	Additional extruded polystyrene insulation board to transformer room	–	5,000.00	8,000.00	Agreed
MC5	Increased the height of toilet cabinet for WC to all typical and disabled toilets	(20,000.00)	35,000.00	18,000.00	Under negotiation
	Structural Instruction				
S1	Additional reinforcement due to revised framing plan and details	–	–	400,000.00	Provisional quantities, assessed in reinforcement re-measurement
S2	Additional works for concrete wall (W15) relocation on 3/F	–	40,500.00	40,250.00	Agreed
S3	BD's approval and consent for amendment of general building plan	(No Cost Effect)		60,000.00	Agreed

Table 6.4 (Continued)

	Building Services Instruction				
BS1	Revision of electrical installation at G/F	(90,000.00)	75,000.00	(15,000.00)	Agreed
BS2	Add 2 No. fresh air grilles to oval F.A. duct at Assembly Hall	–	2,5000.00	40,000.00	Agreed
	Variations without architectural instructions covered (including verbal instruction with the contractor's confirmation letter (CI))				
1	Revised the size of the countertop at 2/F open pantry (Contractor's CI No. 2)	(5,000.00)	6,500.00	2,000.00	Pending architect's instructions
2	New block wall for enclosure of drainage pipe behind 1/F lavatory (Contractor's CI No. 3)	–	2,500.00	Not yet submitted	Pending architect's instructions
	Claims submitted by the contractor				
1	Prolongation cost for 40 days submitted by the contractor (letter dated 6 Aug 2012)	–	160,000.00	500,000.00	Pending architect's advice

9   **Nomination updated information**
   9.1) Nominated sub-contract for lift installation was awarded to Lift Fast Co. Limited in the sum of HK$4,500,000.00.
   9.2) Nominated sub-contract for electrical and HVAC installation was awarded to Elect Good Co. Limited in the sum of HK$8,320,000.00.
   9.3) Nominated sub-contract for fire services installation will be awarded at the end of Sept 2021.

10)   **Status of provisional sum expenditure**
   10.1 The architect instructed the main contractor to carry out the signage works in accordance with the drawings issued under Architect Instruction No. MC6 in the agreed-upon amount of HK$560,000.00.

11)   **Re-measurement status of the provisional quantities**
   11.1 The final set of reinforcement drawings was issued on 1 July 2012. The re-measurement of provisional quantities of reinforcement-has been completed. The total cost of reinforcement is HK$2,500,000.00 instead of HK$1,200,000.00, as allowed in the Bills of Quantities.
   11.2 Other re-measurement for provisional items has not been completed and will be completed in the final measurement period.

Table 6.4 (Continued)

12)   **Prime cost rates in bills of quantities**

Ref:	Description	Qty	Unit	Rate	$	¢	Rate advised under the architect's instruction
	Plasterer						
	Non-slip homogenous floor tiles (P.C. rate $200/m²)						
8/2A	Supply only	20	M²	200.00	4,000.	00	180.00
8/2B	Allow for laying, wastage, profit, etc.	20	M²	155.00	3,100.	00	
	Carpet floor tiles (P.C. rate $350/ m²)						
8/2C	Supply only	50	M²	350.00	17,500.	00	370.00
8/2D	Allow for laying, wastage, profit, etc.	50	M²	300.00	15,000.	00	
	Ironmongery						
	Supplying and fixing ironmongery to door Approved butt hinges; P.C. Supply rates of $40/NO.						
9/3A	Supply only	26	No.	40.00	1,040.	00	60.00
9/3B	Allow for fixing, wastage, profit, etc.	26	No.	20.00	520.	00	
	Approved door handles; P.C. Supply rates of $90/NO.						
9/3C	Supply only	13	No.	90.00	1,170.	00	85.00
9/3D	Allow for fixing, wastage, profit, etc.	13	No.	30	390.	00	

Answer to Example 2: Preparation of financial report

Table 6.5 Answer to Example 2: Preparation of Financial Report

Readers should demonstrate their understanding of how to prepare a financial statement, which should include the following.

1)   Cover page including the (i) project name, (ii) client's name and (iii) date.

2)   Content page

3)   Contents include the pages of (example provided is for demonstration purposes, readers can use different format or templates to address the following contents):

    3.1   Project Financial Summary;

    3.2   Summary of the Financial Statement for the Main Contract;

    3.2   Adjustment for Variations;

    3.4   Adjustment for Anticipated Variations;

    3.5   Adjustment for Prime Cost and Provisional Sums;

    3.6   Adjustment for Provisional Quantities; and

    3.7   Adjustment for Prime Cost Rates.

<u>**Completed set of Financial Report**</u>

Cover Page:

FINANCIAL STATEMENT NO. XX
FOR
ROBINSON INTENERNATIONAL SCHOOL
AT
SAI KUNG, HONG KONG

Accurate Cost Consultant
Date: xx September 2021

P.1

Content Page

CONTENTS

P.2

PROJECT FINANCIAL SUMMARY

1.0	**PROJECT FINANCIAL SUMMARY**	HK$
1.1	Final Contract Sum for Foundation Contract	10,000,000.00
1.2	Anticipated Final Contract Sum for Main Contract (Ref: Appendix A)	80,700,075.00
1.3	Direct Fitting Out Contract (Award Contract Sum)	1,000,000.00
	Total:	91,700,075.00
	<u>Less</u> Budget Allowance	91,000,000.00
1.4	**Anticipated Cost Overrun**	HK$700,075.00

P.3

(Continued)

Table 6.5 (Continued)

MAIN CONTRACT SUMMARY

2.0 Summary of Financial Statement For Main Contract as at xx Sept 2012 (Appendix A)

	Omission (HK$)	Addition (HK$)	HK$
Main Contract Sum (including HK$3,000,000.00 in Contingencies)			80,000,000.00
Less Contingencies			(3,000,000.00)
Adjustment for Variations (Appendix B)	(274,980.00)	353,600.00	
Adjustment for Anticipated Variations (Appendix C)	(5,000.00)	9,000.00	
Adjustment for Prime Cost and Provisional Sums (Appendix D)	(18,190,000.00)	20,006,400.00	
Adjustment for Provisional Quantities (Appendix E)	(1,200,000.00)	2,500,000.00	
Adjustment for Prime Cost Rates (Appendix F)	(23,710.00)	24,765.00	
Anticipated Claims on Prolongation	–	500,000.00	
Total Omission/Addition	(19,693,690.00)	23,393,765.00	
Less Omission		(19,693,690.00)	
Net Addition			3,700,075.00
Anticipated Final Contract Sum (Less Contingencies)			HK$80,700,075.00
Add Remaining Contingencies			0.00
Anticipated Final Contract Sum Including Contingencies			HK$80,700,075.00 =========

P.4

ADJUSTMENT FOR VARIATIONS

<u>**3.0 Adjustment for Variations**</u> (Appendix B)

A.I. No.	Description	Omission (HK$)	Admission (HK$)
	Architectural Instruction		
MC1	Commencement of Contract	(No Cost Effect)	
MC2	Revised architectural drawings at 1/F	(163,000.00)	193,300.00
MC3	Revised aluminum windows (W2)	(1,980.00)	2,300.00
MC4	Additional extruded polystyrene insulation board in transformer room	–	5,000.00
MC5	Increased the height of toilet cabinet for WC to all typical and disabled toilets	(20,000.00)	35,000.00
	Structural Instruction		
S1	Additional reinforcement due to revised framing plan and details	–	–
S2	Additional works for concrete wall (W15) relocation on 3/F	–	40,500.00
S3	BD's approval and consent for amendment of general building plan	(No Cost Effect)	
	Building Services Instruction		
BS1	Revision of electrical installation at G/F	(90,000.00)	75,000.00
BS2	Add 2 No. fresh air grilles to oval F.A. duct at assembly hall	–	2,500.00
	Total Omission/Addition Carried to Summary P.5	(274,980.00)	353,600.00

(Continued)

Table 6.5 (Continued)

ADJUSTMENT FOR ANTICIPATED VARIATIONS

4.0 Adjustment for Anticipated Variations (Appendix C)

A.I. No.	Description	Omission (HK$)	Admission (HK$)
1	Revised the size of countertop at 2/F open pantry (BW's CI No. 2)	(5,000.00)	6,500.00
2	New block wall for enclosure of drainage pipe behind 1/F lavatory (BW's CI No. 3)	–	2,500.00
	Total Omission/Addition Carried to Summary	(5,000.00)	9,000.00
	P.6	=====	====

5.0 Adjustment for Prime Cost and Provisional Sums (Appendix D)

	Description	Omission (HK$)	Admission (HK$)
	Provisional Sums		
1	Fence wall along northern site boundary	(250,000.00)	250,000.00
2	Signage work	(600,000.00)	560,000.00
3	Provide the sum of $4,000,000.00 for lifts installation	(4,000,000.00)	4,500,000.00
	Add __1__ % for profit	(40,000.00)	45,000.00
	Add for attendance as described	(40,000.00)	45,000.00
4	Provide the sum of $7,000,000.00 for electrical and HVAC installation	(7,000,000.00)	8,320,000.00
	Add __1__ % for profit	(70,000.00)	83,200.00
	Add for attendance as described	(70,000.00)	83,200.00
5	Provide the sum of $6,000,000.00 for fire services installation	(6,000,000.00)	6,000,000.00
	Add __1__ % for profit	(60,000.00)	60,000.00
	Add for attendance as described	(60,000.00)	60,000.00
	Total Omission/Addition Carried to Summary	(18,190,000.00)	20,006,400.00
	P.7	========	======

6.0 Adjustment for Provisional Quantities (Appendix E)

Ref.	Description	Omission (HK$)	Admission (HK$)
3/1A-F, 3/2A-H, & 3/3A-G	Reinforcement re-measurement	(1,200,000.00)	2,500,000.00
	Total Omission/Addition Carried to Summary P.8	(1,200,000.00) =======	2,500,000.00 ======

7.0 Adjustment for Prime Cost Rates (Appendix F)

Ref	Description	Omission (HK$)	Admission (HK$)
	Bill No. 8 – Plasterer		
8/2A	Non-slip homogenous floor tiles (P.C. rate $200/m^2)	(4,000.00)	3,600.00
8/2C	Carpet floor tiles (P.C. rate $350/m^2)	(17,500.00)	18,500.00
	Bill No. 9 – Ironmongery		
9/3A	Approved butt hinges; P.C. supply rate of $40/NO.	(1,040.00)	1,560.00
9/3C	Approved door handles; P.C. supply rate of $900/NO.	(1,170.00)	1,105.00
	Total Omission/Addition Carried to Summary P.9	(23,710.00) ======	24,765.00 =====

Final account stage

A final account for a construction contract is the agreed-upon statement of the amount of money to be paid at the end of a construction project by the employer to the contractor for the final settlement of the Contract. According to clause 32.6 of the Contract, the quantity surveyor should prepare the final account, which should include all adjustments (clause 32.7) to be made to the contract sum, including the following:

(i)   provisional work;
(ii)   variations approved by the architect;
(iii) daywork accounts;
(iv)   adjustments for specialist supply and sub-contracting;
(v)   fluctuation costs (if the contract includes a provision for fluctuation); and
(vi)   assessment of contractual claims.

These adjustments must have been made before the completion of the final account as stated in the contract appendix (if not specified, it is assumed to be 12 months), commencing on the date of the substantial completion of the whole works.

The roles of the Quantity Surveyor in preparing the final account are:

(i)   When the contractor agrees to the total sum in the final account, it becomes the total amount that the employer must pay to the contractor for completing all of the works of the contract.
(ii)   The final account is an assessment that confirms the amount stated in the final certificate.
(iii) The details of the final account are subject to inspection by both the contractor and the employer.
(iv)   Usually, the contractor is asked to agree to the final account before the final certificate is issued.

Preparing the final account

A final account typically consists of the following components:

(i)   variation accounts (issued under the architect's instructions or deemed as variations according to the Contract);
(ii)   adjustment of prime cost sums;
(iii) adjustment of provisional sums;
(iv)   adjustment of provisional quantities;
(v)   adjustment of prime cost rates; and
(vi)   claims such as direct losses and expenses.

A sample final account is presented in Example 3.

Table 6.6 Example 3: Sample final account statement

Proposed Residential Development
Main Contract
Statement of Final Account

Main Contractor: ABC Construction Co. Ltd.
Address: 5/F, 123 Nathan Road, XYZ Commercial Centre, Kowloon
Awarded Contract Sum: 108,000,000.00

	OMISSIONS HK$	ADDITIONS HK$
Awarded Contract Sum		108,000,000.00
Omit Contingencies	5,000,000.00	-
Adjustment of Provisional Sums (Appendix A)	3,000,000.00	2,500,000.00
Adjustment of Prime Cost Sums (Appendix B)	22,000,000.00	20,000,000.00
Adjustment of Provisional Quantities (Appendix C)	2,000,000.00	3,000,000.00
Adjustment of Prime Cost Rates (Appendix D)	1,500,000.00	1,300,000.00
Adjustment of Issued Variations (Appendix E)	500,000.00	4,200,000.00
Assessment of Claims		1,000,000.00
(Appendix F)	34,000,000.00	140,000,000.00
Less OMISSIONS		(34,000,000.00)
Final Contract Sum		106,000,000.00

The adjustment processes in Appendixes A to F for the final account are similar to those presented in the financial report in Example 2.

6.7 Chapter summary

This chapter discussed the importance of a cost control and monitoring system for a construction project. It presented (i) the steps and procedures for cost control system, (ii) the steps for preparing a financial statement and (iii) the step for preparing a final account with an example. A detailed example of a project financial statement was also provided.

6.8 Case study

The financial report preparation exercise in Example 2 provides a detailed demonstration of the steps of preparing a financial report for the employer, which is an important step in the construction cost control and monitoring of a construction project.

7   Dispute management

Learning outcomes

Upon completion of this chapter, you should be able to do the following.

1   Appreciate dispute management as an important skill for quantity surveyors.
2   Understand the features of mediation, adjudication and arbitration.
3   Recognise emerging trends in dispute avoidance.
4   Be able to match dispute resolution mechanisms with project particulars.

7.1 Introduction

Managing disputes is one of the core functions of quantity surveyors. Under the Hong Kong Institute of Surveyors' Quantity Surveying Division's assessment of professional competence, understanding various dispute resolution techniques is one of the standards for dispute resolution competency. Being able to practise dispute avoidance and various dispute resolution procedures is also required to reach Level-2 (Doing) in the Royal Institution of Chartered Surveyors' assessment of competence guidelines.

From a broader perspective, Hong Kong aspires to become a hub for dispute resolution services in the Asia-Pacific region. Paragraph 31 of the 2014 Hong Kong Policy Address (HKSAR, 2014) states the following:

> Hong Kong has a fine tradition of the rule of law and a well-developed legal system. The Government will continue to actively promote Hong Kong's legal and dispute resolution services to enhance our status as a centre for international legal and dispute resolution services in the Asia-Pacific region. The Government will strengthen its promotion efforts overseas, continue to co-ordinate the development of mediation services through the Steering Committee on Mediation, and establish an advisor committee on the development and promotion of arbitration services.

Further development of this policy decision is found in the 2015 Hong Kong Policy Address (HKSAR, 2015):

> 51 Hong Kong's tradition of rule of law and our legal system are conducive to the development of legal and dispute resolution services. Last year, the China Maritime Arbitration Commission established an office in Hong Kong, its first branch office outside the Mainland. The Central Government and the HKSAR Government have recently signed

DOI: 10.1201/9781003212355-7

the Host Country Agreement and the related Memorandum of Administrative Arrangements respectively with the Permanent Court of Arbitration on the conduct of dispute settlement proceedings in Hong Kong by the Court.

52   Renowned arbitration institutions such as the International Court of Arbitration of the International Chamber of Commerce, the China International Economic and Trade Arbitration Commission have set up offices in Hong Kong in recent years. With increasing maritime activities in Asia, maritime arbitration services have immense potential for growth. Arbitration awards made in Hong Kong are enforceable in the jurisdictions of over 150 contracting states of the New York Convention, and also in the Mainland and Macao. We will actively further advance development in this area.

According to the Hong Kong International Arbitration Centre (2019), Hong Kong is uniquely positioned to be a regional dispute resolution centre for the following reasons.

1   Independent and neutral forum
Hong Kong upholds the rule of law through its common law legal system, which is overseen by an independent judiciary comprised of local and international judges who are independent, professional and efficient. Hong Kong is ranked fourth out of 148 countries on the index of judicial independence published in the World Economic Forum's Global Competitiveness Report 2013–2014. Parties are free to choose lawyers and arbitrators from anywhere in the world without restriction.
2   World class legislative framework and arbitration-friendly judiciary
The HK Arbitration Ordinance (CAP609) enacted on 1 June 2011 affirms Hong Kong's status as a user-friendly United Nations Commission on International Trade Law (UNCITRAL) jurisdiction. The CAP609 provides explicit assurance of confidentiality in arbitral proceedings, awards and related court proceedings. Hong Kong's courts are also pro-arbitration.
3   Professional services
Hong Kong has an extraordinarily large pool of multilingual professionals. As of early 2020, its pool of professionals included over 1,100 barristers, 94 of whom were senior counsel, over 6,700 local practicing lawyers and 1,500 registered foreign lawyers, 29,000 engineers, 37,000 accountants, more than 8,500 members of the Hong Kong Institute of Surveyors and over 4,000 architects according to the Hong Kong Institute of Architects.

Professional quantity surveyors with dispute resolution expertise are thus ideally placed to contribute to Hong Kong's development into a dispute resolution services hub.

7.2 Identification of construction disputes

A dispute resolution clause is now a regular provision in construction contracts. Cheung and Pang (2013) explored the anatomy of construction contracts from a functional perspective. Figure 7.1 displays the functions of a construction contract. The centre indicates the employer and the contractor's main obligations. The next ring, adjustment, includes provisions to deal with variations and the associated time and financial compensation. The control and approval ring covers the measures that allow an employer to exercise control, such as

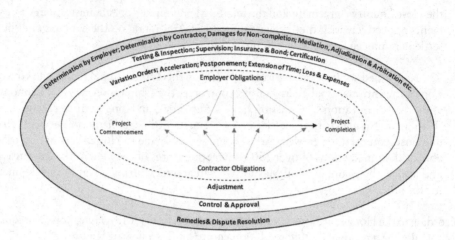

Figure 7.1 Functional analysis of construction contracts

Source: Adapted from Cheung and Pang (2013)

testing and inspection. Certificates are used to signify approval. The outermost ring includes contractual remedies and dispute resolution arrangements.

A contract's conditions usually define what is considered a dispute. For example, under clause 86(1) of the 1999 edition of the Hong Kong Government General Conditions of Building Contract (HKSAR,1999), disputes are identified as follows:

> If any dispute or difference of any kind whatsoever shall arise between the Employer and the Contractor in connection with or arising out of the Contract or the carrying out of the Works including any dispute as to any decision, instruction, order, direction, certificate of the Architect or certificate or valuation by the Surveyor whether during the progress of the Works or after their completion and whether before or after the termination, abandonment or breach of the Contract, it shall be referred to and settled by the Architect who shall state his decision in writing and give notice of the same to the Employer and the Contractor.
>
> (Emphasis added)

The scope of disputes is very wide, as any difference between the employer and the contractor can trigger the dispute resolution procedure. Similar wording is used in the 1986 Standard Form of Building Contract published by the Hong Kong Institute of Architects, the Royal Institution of Chartered Surveyors (Hong Kong Branch) and the Society of Builders. Moreover, the 2005 Agreement and Schedule of Conditions of Building Contract published by the Hong Kong Institute of Architects, the Hong Kong Institute of Construction Managers and the Hong Kong Institute of Surveyors does not list conditions for where and when disputes could arise; the settlement of disputes clause (clause 41) instead sets out the procedures to be applied to settle any disputes that arise.

The escalation of differences into disputes may be driven by various factors. Mururu (1991) defined disputes as the formation of a position to be maintained in a conflict, whereas Brown and Marriot (1999) proposed that disputes should be regarded as conflicts that require resolution. Construction disputes are often complex in nature and can involve numerous stakeholders, which increases the difficulty of attaining a universally accepted definition. Spittler and Jentzen (1992) argued that construction disputes are associated with differences in parties' perspectives, interests and agendas. A high number of stakeholders can exacerbate a dispute. Tillett and French (1991) observed that construction disputes are caused by the incompatibility of two or more groups' interests, needs or goals.

Construction disputes have been extensively researched. For example, Diekmann et al. (1994) classified the causes of construction disputes into three types: people, processes and products. Rhys Jones (1994) identified 10 causes of disputes: (1) management, (2) culture, (3) communication, (4) design, (5) economics, (6) tendering pressures, (7) law, (8) unrealistic expectations, (9) contracts and (10) workmanship. Hewitt (1986) classified construction disputes into six groups: (1) change of scope, (2) change of conditions, (3) delays, (4) disruptions, (5) acceleration and (6) terminations. Disputes can be further broken down into 59 different subtypes belonging to six general types: (1) determination of contracts, (2) payment related, (3) the site and execution of work, (4) time related, (5) final certificate and final payment and (6) tort-related. Heath et al. (1994) identified seven major types of construction disputes: (1) contract terms, (2) payments, (3) variations, (4) extensions of time, (5) nominations, (6) re-nominations and (7) availability of information. Conlin et al. (1996) identified six causes of construction disputes: (1) payment, (2) performance, (3) delay, (4) negligence, (5) quality and (6) administration. Kumaraswamy (1997) further ranked the causes based on the frequency and magnitude of the disputes: (1) variations due to site conditions, (2) variations due to client changes, (3) variations due to design errors, (4) unforeseen ground conditions, (5) ambiguities in contract documents, (6) variations due to eternal events, (7) interferences with utility lines, (8) exceptional inclement weather, (9) delayed design information and (10) delayed site possession.

Disputes that trigger the dispute resolution clause are deemed contractual disputes. Semple et al. (1994) argued that site overhead, loss in productivity or revenue and financing costs are the major dispute types in construction contracts. Yates (1998) proposed seven causes of contractual disputes: (1) variations, (2) ambiguities in contract documents, (3) inclement weather, (4) late issue of design information/drawings, (5) delayed possession of site, (6) delay by other contractors employed by the developer and (7) postponement of part of the project.

In Hong Kong, the most common dispute issues settled by mediations are variation, progress delays, the expectations of involved parties and problems arising between parties (Yiu and Cheung, 2004). Spittler and Jentzen (1992) argued that ambiguity in contract documents, adversarial attitudes and perceptions of unfairness cause most disputes. The contractual provisions themselves can also lead to disputes (Semple et al., 1994). Semple et al. (1994) proposed that the two major sources of construction disputes are contracts and unpredictable events. To prevent disputes caused by uncertainties, contradictory contractual provisions must be avoided. Applying the transaction cost framework proposed by Williamson (1979), Mitropoulos and Howell (2001) proposed that the fundamental factors causing disputes were (1) project uncertainty, (2) contractual problems and (3) opportunistic behaviours.

In summary, construction disputes can be viewed from a number of perspectives. The most common approaches focus on subject matter, but others highlight the human factors. Thus, the appropriate dispute resolution procedure may depend on the dispute's characteristics. Section 7.3 discusses three commonly used construction dispute resolution mechanisms.

7.3 Commonly used dispute resolution mechanisms

7.3.1 Mediation

The Mediation Ordinance (CAP 620) was enacted in 2013 after the Hong Kong Government decided to develop Hong Kong into a dispute resolution services centre for the Asia-Pacific Region. The objectives of CAP 620 are (i) to promote, encourage and facilitate the resolution of disputes through mediation and (ii) to protect the confidential nature of mediation communication.

Section 4 of CAP 620 describes mediation as a structured process comprised of one or more sessions in which one or more impartial individuals, without adjudicating a dispute or any aspect of it, assist the parties to the dispute in doing any or all of the following:

1   identifying the issues in dispute;
2   exploring and generating options;
3   communicating with one another; and
4   reaching an agreement regarding the resolution of the whole, or part, of the dispute.

Mediation has become an integral part of the construction dispute resolution regime because almost all commonly used construction contracts have incorporated mediation into their dispute resolution procedures. Mediation is a voluntary process though which an impartial individual (mediator) helps the disputing parties to resolve their disputes. It can thus be said that mediation is a form of assisted negotiation. The assistance of a mediator is therefore pivotal. Table 7.1 presents the key stages of a typical mediation and the objectives and strategies of the mediator.

The distinctive features of mediation include voluntary participation and confidential proceedings. An agreement reached via voluntary participation is most likely to be honoured. An agreement reached through mediation becomes a contract between the disputing parties. Private proceedings also prevent disputes from becoming public.

In addition to the popular contractual use of mediation for construction disputes, the high court Practice Direction 6.1 ('PD 6.1') (HKHC, 2009) indirectly promotes the use of mediation for construction disputes before court hearings.

Table 7.1 Key stages of a mediation

Stages	Objectives	Goals of the Mediator
Introduction	Explain the role of the mediator and the process	Ice-breaking and trust-building
Opening Joint Session	Gather information and positions; ensure that parties are heard	Solicit wish list
Initial Caucuses	Gather information and positions beyond those in the opening joint session	Identify important issues, especially those requiring reconciliation
Subsequential Caucuses	Explore options and move for agreement	Shuttle diplomacy; reality testing and upholding confidentiality
Closing Joint Session	Ensure that agreement has been reached	Commitment to the agreement

Section F of PD 6.1 reiterates the voluntary nature of mediation and encourages parties to participate. The relevant PD 6.1 provisions are as follows.

20 Parties in construction cases are <u>encouraged to attempt</u> mediation as a possible cost-effective means of resolving disputes.

21 To promote the use of mediation, the <u>Court may impose cost sanctions</u> where a party <u>unreasonably refuses</u> to attempt mediation.

23 A party ('the Applicant') to a construction action may serve a <u>Mediation Notice</u> upon any other party ('the Respondent') in the dispute. A copy of the Mediation Notice is to be filed with the Court.

26 The Mediation Notice should specify <u>a timetable</u> for the proposed mediation, stating what <u>minimum amount of participation</u> by the Respondent would qualify (as far as the Applicant was concerned) as a sufficient attempt at mediation.

28 <u>Where the Respondent refuses to mediate the relevant dispute, the Respondent should also state why it does not believe that mediation is appropriate.</u>

37 No party to an action shall be compelled to go to mediation. The Court will treat an agreement to mediate reached pursuant to Section F (2) above ('a Mediation Agreement') as having been arrived at on a <u>purely voluntary</u> basis, without prejudice to the parties' contentions in the action.

40 The conduct of the mediation shall remain <u>confidential</u> to the parties and will proceed wholly on a without prejudice basis.

41 <u>Where a Mediation Notice has been served, an unreasonable refusal or failure to attempt mediation may expose a party to an adverse costs order.</u>

42 Where a party:
(1) <u>has engaged in mediation up to the minimum level of expected participation</u> agreed by the parties beforehand or as determined by the Court; or
(2) has <u>a reasonable explanation for non-participation,</u>
he should not suffer any adverse costs order.

43 What constitutes an adverse costs order will be a matter in the Court's discretion after taking into account all relevant circumstances.

44 In determining whether a party has acted unreasonably in refusing mediation, the Court <u>will not</u> take account of or inquire into:
(1) what happened during the mediation;
(2) why the mediation failed; or
(3) whether any failure in the course of mediation may be ascribed to unreasonable conduct by any party.

(Emphasis added)

PD 6.1 is applicable to cases reaching the High Court Construction and Arbitration List, but its two distinctive features, voluntary participation and confidential proceedings, are maintained to encourage attempts at mediation. However, the use of the adverse costs order makes the participation less voluntary.

7.3.2 Adjudication

Adjudication has seen relatively little use in Hong Kong, although it was incorporated into the 1992 Conditions of Contract for the Airport Core Programs, where payment and extension

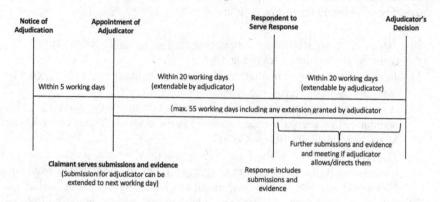

Figure 7.2 Adjudication timetable and procedure

Source: Adapted from DEVB (2015)

of time-related disputes were handled by adjudication. In 2015, the Hong Kong Government proposed a security of payment legislation (SOPL) in line with similar legislation passed in the United Kingdom, Australia, New Zealand, Singapore and Malaysia. The primary aim of the Hong Kong SOPL is to ensure that contractors, sub-contractors, consultants and suppliers are paid in a timely manner. In addition, controversial pay-when-paid or pay-if-paid clauses are prohibited. Figure 7.2 shows the timetable and procedure of adjudication under SOPL.

Chapter 6 of the SOPL consultation paper details the proposed applications of adjudication, as follows (DEVB, 2015):

Disputes are referred to an adjudicator who will <u>independently consider and decide the dispute by issuing a written decision</u> which will set out any amount to be paid. If either party is unhappy with the decision, they can <u>take the dispute to court (or arbitration if the contract provides for it) for a final determination in the usual way</u>. This would not be an appeal of the adjudicator's decision in court proceedings or arbitration as stipulated in the contract.

The adjudicator's decision is <u>binding and enforceable on an interim basis</u> in the same way as a court judgment and there can be no set off against an adjudicator's decision. Even if one of the parties takes the dispute on to court or arbitration, payment of the adjudicator's decision must be made in the meantime. When court or arbitration proceedings are concluded, it may be that further amounts have to be paid or repaid.

<u>Courts can enforce adjudicator's decisions even if it is apparent that they are legally or factually wrong.</u> This is because the overriding objective of adjudication is to provide a rapid independent decision. It is inevitable that decisions may at times be rough and ready but parties can still go to court or arbitration for a final considered judgment or award which will be given after completion of the full litigation or arbitration process.

<u>The right to adjudication cannot be contracted out of and cannot be limited.</u> For example, contractual provisions seeking to limit and define when a dispute is considered to arise for the purposes of adjudication will be ineffective. Also, provisions requiring an Engineer's/Architect's decision on the dispute or similar as a precondition to adjudication will be ineffective. Adjudicators will be able to review certified amounts relied on for the purposes of Payment Responses and decide that different amounts are due.

Both parties to a contract will be entitled to refer disputes to adjudication but <u>limited to disputes</u> concerning the following:

a)   the value of work, services, materials and plant supplied and claimed in a Payment Claim; and/or

b)   other money claims made in accordance with any provision of the contract and claimed in a Payment Claim; and/or

c)   set offs and deductions against amounts due under Payment Claims; and/or

d)   the time for performance or entitlement to extension of the time for performance of work or services or supply of materials or plant under the contract.

(Emphasis added)

Adjudication can alter the power asymmetry between contracting parties. Without the right to adjudication, a payment-related dispute may take months or years to resolve, and the party with fewer resources may be disadvantaged. An unpaid party facing a costly and protracted dispute is likely to yield to pressure. A relatively simpler resolution procedure such as adjudication provides a less expensive way to resolve the dispute, although the adjudication may produce interim results.

7.3.3 Arbitration

In Hong Kong, arbitration is a regulated process. Governing laws include the common law, CAP609 and the relevant arbitration rules. Arbitration was once considered an alternative to litigation, but the legalistic approach required has made the arbitration process more similar to civil litigation. Section 3 of CAP609 (HKSAR, 2010) sets out its objectives and principles.

1   The object of this Ordinance is to facilitate <u>the fair and speedy resolution</u> of disputes by arbitration without <u>unnecessary expense</u>.

2   This Ordinance is based on the principles

(a)   that, subject to the observance of the safeguards that are necessary in the public interest, the parties to a dispute should be <u>free to agree</u> on how the dispute should be resolved; and

(b)   that the court should <u>interfere</u> in the arbitration of a dispute <u>only as expressly provided for</u> in this Ordinance.

(Emphasis added)

In arbitration, the principles of natural justice should be observed, such that

1   each party has an equal and reasonable opportunity to present his/her case and address that of his/her opponent; and

2   the adjudicator acts impartially.

To operationalise these principles, the following guidelines have been developed.

1   Each party must have notice of any meeting, hearing or determination of a matter in the arbitration.

2   Each party must have a reasonable opportunity to be present and to be heard at such meeting, hearing or determination, together with his/her advisers and witnesses.

3   Each party must have a reasonable opportunity to be present throughout such a meeting, hearing or determination to present evidence and arguments in support of his/her own case.

4   Each party must have a reasonable opportunity to test his/her opponent's case by cross-examining his/her witnesses, presenting rebutting evidence and addressing oral arguments.

5   The hearing must, unless the contrary is expressly agreed upon, be the occasion on which the parties present the whole of their evidence and arguments.

6   The arbitrator may only decide the case on the evidence and submissions placed before him/her by the parties.

7   If one party requests an oral hearing, an arbitrator is not required to accede to such a request. He/she must rather adopt a procedure that is appropriate given the circumstances of the case, avoiding unnecessary delay and expense, to provide a fair means of resolving the dispute.

Arbitration is typically the last resort of contractual disputes. Under CAP609, it is nearly impossible to skip arbitration and proceed with litigation. Upon the completion of the necessary mediation and/or adjudication procedures, a dispute can be referred to arbitration. Figure 7.3 shows the typical flow of arbitration.

The arbitration procedure starts with the claimant referring a dispute for arbitration. The arbitrator is usually jointly appointed by the claimant and the respondent. If agreement cannot be reached on the arbitrator, a nomination is made by an independent organisation as stipulated in the contract. For example, under the 2005 Agreement and Schedule of Conditions of Building Contract, the arbitrator shall, upon request by either the employer or the contractor, be appointed by the President or Vice President of the Hong Kong Institute of Architects co-jointly with the President or Vice President of the Hong Kong Institute of Surveyors (HKIA et al., 2005).

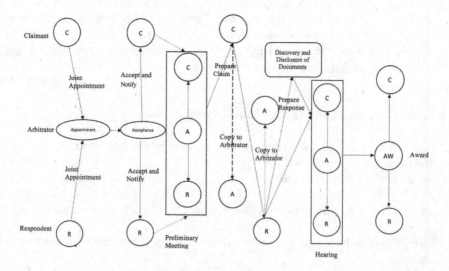

Figure 7.3 Flow chart of a typical arbitration

After the arbitrator has accepted the appointment, a preliminary meeting is organised to agree upon the administrative matters of the proceeding, such as the programme for submissions, scope of discovery, arrangement of mail and date. The claimant then prepares their claim in accordance with the agreed-upon timeline. Likewise, the respondent provides their responses and counterclaim, if any. There may be further exchanges of documents for the claim, responses and counterclaim. It is also common for parties to request further and better particulars for these documents. Discovery and disclosure of documents is also carried out. The parties must list the documents that are or have been in their possession, control or custody. The parties can thus review their cases in the light of what the counterpart may have. The parties will present their cases at a hearing, with witnesses of fact and expert witnesses involved. In most arbitration cases, the parties are legally represented. Upon completion of the hearing, the arbitrator announces his/her award in due course.

An arbitrator's power is wide-ranging. For example, under clause 41.6 of the 2005 Agreement and Schedule of Conditions of Building Contract (HKIA et al., 2005), the arbitrator's powers include the following:

1   rectifying the contract to accurately reflect the true agreement made by the parties;
2   directing measurements or valuations to determine the rights of the parties;
3   assessing and awarding any sum that ought to have been the subject of or included in a certificate; and
4   opening up, reviewing and revising, without limitation, the giving, submitting or issuing of any agreement, approval, assessment, authorisation, certificate, confirmation, consent, decision, delegation, direction, dissent, determination, endorsement, instruction, notice, notification, opinion, request, requirement, statement, termination or valuation.

7.4 Dispute avoidance

7.4.1 *From resolution to avoidance*

Industry reviews conducted both overseas and in Hong Kong have identified inefficient dispute resolution as an issue that merits serious attention. For example, recommendation number 61 of the Construction Industry Review Committee (CIRC) report (CIRC, 2001) urged the adoption of suitable procurement to facilitate integration of the project team. Furthermore, the industry has been urged to improve efficiency, as a more integrated project team can offer a higher standard of service and fewer disputes. Since the release of the CIRC report in 2001, there has been an increase in the use of partnering.

7.4.2 *The ASD DRAd system*

In the 1980s, the Architectural Services Department introduced the Dispute Resolution Advisor System (DRAd) (Cheung and Yeung, 1998), and in August 2019, an updated version was released. The new version is specified in the Special Conditions of Contract clause 59 (SCC 59 hereafter) to the Hong Kong Government General Conditions of Contract 1999 edition (HKGGCC) (HKSAR, 2019).

Aim of the DRAd system (HKSAR, 2019)

To foster cooperation between the Employer and the Contractor and among Specialist Sub-Contractors and Nominated Sub-contractors who may be engaged for the execution of any part of the works to <u>minimize</u> the volume of claims, disputes and disruptions to the works, and to ensure the <u>cost-effective</u> and <u>expeditious resolution of these disputes</u> that do arise. (clause 59 (1))

Appointment of DRAd

A DRAd shall be appointed within a maximum of <u>60 days</u> of the award of the contract. (clause 59 (3)(a))

Should the Employer and the Contractor fail to agree on a suitable DRAd within the specified period, the DRAd shall be appointed upon the written application of either the Employer or the Contractor by the <u>Hong Kong International Arbitration Centre</u>. (clause 59(3)(e))

Dispute avoidance

The DRAd shall meet on the <u>monthly</u> basis with the Employer and the Contractor either separately or together to attempt to resolve problems that arise <u>before they become formal disputes</u> and to anticipate <u>problems that may arise</u> in the future. (clause 59(4)(6))

According to the Dispute Resolution Advisor (DRAd) System Handbook, Part I, item 1.1.3, the minimum mandatory criteria for the use of the DRAd system (HKSAR, 2019) include the following.

The nature of work should be complicated that disputes are likely to arise during the course of contract (item 1.1.3.2(a)), and

The contract value should be over HK$500M, or over HK$100M for exceptional case where there is demonstrable benefit to adopt DRAd System. (item 1.1.3.2(b))

Typically, the Employer and the Contractor must each pay 50% of the DRAd's cost, fees and expenses for its services (HKSAR, 2019).

After either the Employer or the Contractor serve a <u>Notice of Dispute</u>, the DRAd shall promptly meet with the <u>site level representatives</u> of the Employer and the Contractor, and if appropriate, with the representatives of any relevant specialist sub-contractor or nominated sub-contractor, to attempt to reduce the dispute at site level. The DRAd shall have access to those records that are material to the dispute. The DRAd shall have flexibility in the choice of dispute resolution approach that he can use to help settle the dispute. (clause 59(5)(c))

If the dispute cannot be resolved at site level within 14 days of the service of the Notice of Dispute, the DRAd shall within <u>3 days</u> from the expiry of that <u>14-day</u> period send a written report (hereafter called 'the Report') to non-site senior officers of the Employer and the Contractor. The Report shall be concise and shall analyse the dispute. It shall identify the key issues in dispute and the perceptions of the DRAd as to the <u>obstacles</u> to settlement. If requested in writing by <u>both</u> the Employer and the Contractor, the DRAd shall within 3 days from such request also provide either a <u>non-binding recommendation</u> for a resolution or a <u>non-binding evaluation</u> of the merits of the dispute. The DRAd may show a draft of the Report to the non-site senior officers. The Report of the DRAd shall <u>not</u> be <u>admissible</u> in any subsequent arbitration or litigation except as set forth in sub-clause 59(8) of this clause. (clause 59(5)(e)(i))

Within <u>7 days</u> upon receipt of the Report from the DRAd, the non-site senior officers shall meet in order to attempt to settle the dispute. The DRAd shall only attend this meeting if the Employer and the Contractor request in writing that he do so. (clause 59(5)(e)(i))

If the dispute is not settled within <u>21 days</u> of the date of the service of the Report to the non-site senior officers, the DRAd may recommend to the Employer and the Contractor <u>another form of dispute resolution that</u> in his judgement may be a more effective means of resolving the dispute than the short-form arbitration described in sub-clause (6) of this clause. (clause 59(5)(f)(i))

Short-form arbitration

The short-form arbitration shall be governed by the 'Short-Form Arbitration Rules' set out in the Architectural Services Departments' Standard Form.

(clause 59(6))

The hearing shall involve one or, with the written agreement of the Employer and the Contractor, at most, a limited number of distinct claims or issues.

(clause 59(6)(a))

If the hearing involves one claim or issue, the arbitration shall be conducted and concluded in, at most, <u>one day</u>. If it includes more than one distinct claim or issue, the Employer and the Contractor shall agree on a maximum length of time for the hearing, which shall be as short as possible.

(clause 59(6)(b))

The Report of the DRAd, as well as any prior expressions of his views, shall not be admissible in the arbitration, except as set forth in sub-clause (8) of this clause.

(clause 59(7))

The DRAd Report may be admissible only if it is relevant to the issue of whether the costs of the DRAd involvement with respect to a particular dispute may be transferred from the Employer to the Contractor or from the Contractor to the Employer as the case may be ... The DRAd Report may be admissible only after the arbitration award, save for the award as to cost, has been made and published and then only if the arbitrator believes that it is appropriate to take evidence as to whether costs should be transferred.

(clause 59(8))
(Emphasis added)

7.4.3 *The HKHA dispute avoidance and resolution advisor system*

The Hong Kong Housing Authority (HKHA) introduced a Dispute Avoidance and Resolution Advisor (DARA) under the 2013 HKHA General Conditions of Contract for Building Works. Figure 7.5 shows a flowchart of the dispute avoidance and resolution procedure. The 2013 HKHA General Conditions of Contract for Building Works (HKHAGCC) extended the function of DRA to DARA.

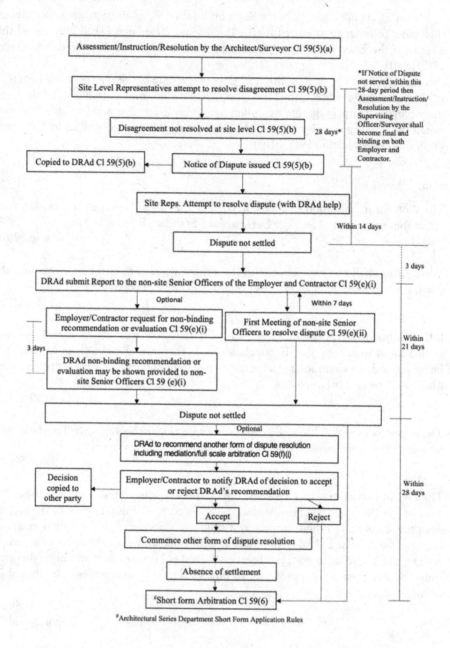

Assessment/Instruction/Resolution by the Architect/Surveyor Cl 59(5)(a)

Site Level Representatives attempt to resolve disagreement Cl 59(5)(b)

Disagreement not resolved at site level Cl 59(5)(b) | 28 days*

*If Notice of Dispute not served within this 28-day period then Assessment/Instruction/Resolution by the Supervising Officer/Surveyor shall become final and binding on both Employer and Contractor.

Copied to DRAd Cl 59(5)(b) ◀— Notice of Dispute issued Cl 59(5)(b)

Site Reps. Attempt to resolve dispute (with DRAd help)

Within 14 days

Dispute not settled

3 days

DRAd submit Report to the non-site Senior Officers of the Employer and Contractor Cl 59(e)(i)

Optional

Employer/Contractor request for non-binding recommendation or evaluation Cl 59(e)(i)

Within 7 days

First Meeting of non-site Senior Officers to resolve dispute Cl 59(e)(ii)

3 days

DRAd non-binding recommendation or evaluation may be shown provided to non-site Senior Officers Cl 59 (e)(i)

Within 21 days

Dispute not settled

Optional

DRAd to recommend another form of dispute resolution including mediation/full scale arbitration Cl 59(f)(i)

Decision copied to other party ◀— Employer/Contractor to notify DRAd of decision to accept or reject DRAd's recommendation

Within 28 days

Accept   Reject

Commence other form of dispute resolution

Absence of settlement

#Short form Arbitration Cl 59(6)

#Architectural Series Department Short Form Application Rules

Figure 7.4 Flowchart of the dispute resolution process under SCC 59 of the HKGGCC
Source: Adapted from the SCC 59 Guidebook

Salient features

There are certain similarities between clause 17 (Avoidance and Resolution of Disputes) of the HKHAGCC and SCC 59 of the HKGGCC. Clause 17 is more explicit in expressing the aims of dispute avoidance. Its notable features are as follows (HKHA, 2017a).

The use of a contract manager who is not a member of the project team will minimise the desirable effect of project team members deciding the dispute. Clause 2.1 details the role of the contract manager (HKHA, 2017b).

> Contract Manager means the person, company or firm named in the Contract as the Contract Manager or such other person, company or firm as may be appointed from time to time by the Employer and notified in writing to the Contractor to act as the Contract Manager for the purposes of the Contract. The person named or appointed may be described by name or as the holder for the time being of a public office.

> The Contract Manager should carry out the duties as are specified in or necessarily to be implied for the contract with due expedition.

> Dispute avoidance and resolution advisor

The following is summarised from clause 17.1 of the HKHAGCC (HKHA, 2017b).

The employer and the contractor jointly select and appoint a DARA in accordance with the specified procedures as soon as is practicable after the issue of the letter of acceptance. The DARA fee is to be shared equally between the parties involved in the dispute.

If a difference arises over a decision regarding the extension of time, certificate of completion, compensation for suspension of the works progress claims or determination, the site-level representative of the employer and the contractor shall attempt to resolve it within 28 days of the date of issue of the stated matters.

If a difference arises over other matters, the employer or the contractor shall notify the other in writing. Within 28 days from the date of such notice, either party may write to the contract manager, sending a copy to the other party, requesting that the contract manager make a decision.

The contract manager should respond to such a request within 14 days with either a decision or a statement of inability to make a decision.

If the difference remains unresolved and either party wishes to pursue resolution, this party shall serve in writing to the other a Notice of Dispute (NOD), with a copy sent to the DARA.

Upon receipt of the NOD, the DARA will meet with the site-level representatives of the disputing parties and help resolve the dispute. If the dispute remains unresolved after 14 days from the serving of the NOD, the DARA will provide a DARA report to the senior management of the disputing parties.

Upon receipt of the DARA report, the senior management of the disputing parties will meet and attempt to settle the dispute. If the effort fails within 14 days of the date of issue of the DARA report, the DARA may propose that ADR be adopted.

If neither party is willing to proceed with the proposed ADR or if the ADR does not lead to a settlement, the dispute can be referred to short-form arbitration within 35 days of the DARA report.

If either party rejects the use of short-form arbitration, the dispute can be referred to arbitration within 28 days of written refusal.

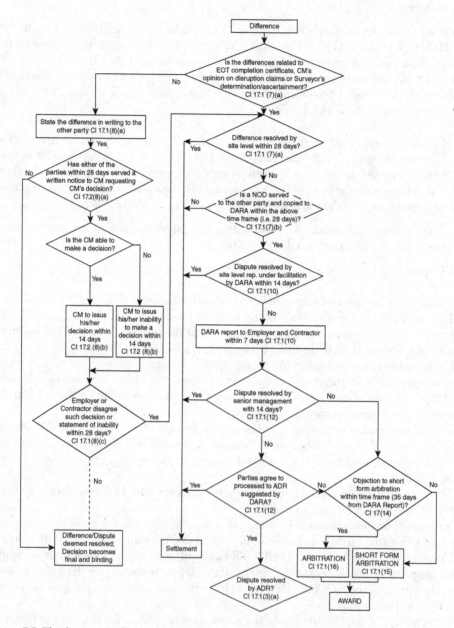

Figure 7.5 The dispute avoidance and resolution procedure under clause 17 of the HKHAGCC
Source: Adapted from the HKHAGCC Guidebook

7.4.4 Summary

The goal of using DRAd and DARA is to resolve problems as early as possible. Advice from a neutral third party (DRAd or DARA) may assist the parties in proactively addressing problems before they become disruptive disputes. In addition, unresolved problems can make disputes more complex, making resolution more difficult. The early involvement of senior members of the contracting parties can help to relieve the tension between site representatives who need to maintain a working relationship. Short-form arbitration can also further reduce the resource burden should the services of an arbitrator become necessary.

7.5 Chapter summary

A dispute resolution clause is included in most construction contracts to resolve differences between contracting parties. Such differences typically have time and cost implications. A contractual dispute resolution provision can also serve a gap-filling function to deal with issues that do not have a planned recourse. This chapter is titled 'Dispute Management' rather than 'Dispute Resolution' because it is more efficient to prevent disputes than to resolve them. Thus, in some contracts, the term 'problem-solving' is used.

This chapter provided an overview of commonly used construction dispute resolution mechanisms, which include mediation, adjudication and arbitration. In recent years, greater efforts have been made to promote dispute avoidance. Two notable systems for applying this concept are the dispute resolution advisor system initiated by the Architectural Services Department and the dispute avoidance and dispute resolution advisor put forward by the Hong Kong Housing Authority. There is no fixed formula for managing disputes in construction projects; a mechanism should be chosen that is appropriate for the issue in dispute. The case discussion in Section 7.6 provides an illustration.

7.6 Case study

7.6.1 Choice of dispute resolution mechanism

This chapter introduced a number of dispute resolution mechanisms commonly used in Hong Kong. Several factors may influence the choice of dispute resolution mechanism for a particular case. First, the sophistication of the parties can be pivotal. For example, most users would opt for mechanisms that have been successful for them in the past. In addition, a party with more resources may prefer more formal proceedings. The nature of the dispute is also critical: the court is considered the best forum for a legal dispute, and where international parties are involved, the value of obtaining an outcome that is internationally recognised cannot be overlooked. The relative rankings of selection criteria found in Cheung (1999) are (i) speed, (ii) economy, (iii) relationship, (iv) flexibility, (v) confidentiality, (vi) enforceability, (vii) privacy, (viii) fairness, (ix) bindingness, (x) control by parties, (xi) width of remedy and (xii) creative solution. Most users are pragmatic and prefer methods that offer tangible and immediate advantages.

7.6.2 A hypothetical case

Background

A developer approaches you for advice on the choice of a resolution process for a major development in Tung Chung. The project has four phases to be completed over 10 years.

The works include major site formation and roadwork, a shopping centre that will be the largest on Lantau Island, 15 high-rise blocks and 50 villas. The development will showcase the developer's smart buildings. Extensive energy-saving construction will be incorporated; as such, state-of-the-art green construction will be used to the extent possible. The developer would like to reserve his/her final choice of the type of green technology until as close to the time of physical construction as possible. Early engagement of specialist sub-contractors and suppliers is also anticipated to tap into their expert knowledge of green technologies. Moreover, time is of the essence, as the whole project has been earmarked for completion by 8 December 2028 (about eight years from the time of awarding the first phase of the project), the 50th anniversary of the owner's marriage.

Discussion

This is a fairly complex development that includes a wide variety of work and spans a period of eight years. There will be many interfacing issues, and extensive use of green technology is also envisaged.

Given the complexity of the operation, both technical and contractual problems are very likely as the works proceed. The completion of the whole development has been earmarked for a specific date to serve as the celebratory event for the owner.

With the above in mind, disputes should be avoided to the extent possible. The use of a standing neutral third party to assist in solving problems that may arise should be considered, with the ultimate aim of minimising major disputes. Reference can be made to clause 17 of the HKHAGCC and SCC 59 of the HKGGCC in this regard. The goal is to solve any problems as quickly as possible.

The inclusion of an alternative dispute resolution process should also be considered for the next tier of resolution. It is recommended that the senior management of the stakeholders be well informed. If a dispute cannot be resolved at this tier, arbitration during the contract period is recommended.

A certain degree of flexibility should be maintained regarding the alternative dispute resolution method. For example, disputes arising from the use of state-of-the-art green technology may be better handled by an expert in the field. The use of expert determination would thus seem appropriate.

In short, the scale and complexity of the project warrant a more collaborative approach, and the relatively tight time frame would be better served using less formal mechanisms. A multi-tiered dispute avoidance and resolution procedure that involves a standing neutral third party, private alternative dispute resolution processes and short-form arbitration is therefore proposed for consideration by the developer.

References

Arcadis (2020a) *Practice Guide Book*, Arcadis Hong Kong Limited.

Arcadis (2020b) *Construction Cost Handbook, China & Hong Kong*, Arcadis Hong Kong Limited.

Brown, H. J., & Marriot, A. L. (1999) ADR: *Principles and Practices* (2nd ed.), Sweet & Maxwell.

Building (Planning) Regulations (2021)

Building Department (2011) Code of practice for authorized persons. Available at: https://www.bd.gov.hk/doc/en/resources/codes-and-references/practice-notes-and-circular-letters/pnap/APP/APP007.pdf

Cheung, S. O. (1999) Critical factors affecting the use of alternative dispute resolution processes in construction, *The International Journal of Project Management*, 17(3):189–194.

Cheung, S. O., & Pang, K. H. Y. (2013) Anatomy of construction disputes, *Journal of Construction Engineering and Management*, 139(January):15–23. doi: 10.1061/(ASCE)CO.1943-7862.0000532.

Cheung, S. O., & Yeung, Y. W. (1998) The effectiveness of the dispute resolution advisor system: A critical review, *The International Journal of Project Management*, 16(6):367–374.

CIRC (2001) *Construct for Excellence: Report of the Construction Industry Review Committee*, The Government of the HKSAR.

Conlin, J., Langford, D., & Kennedy, P. (1996) The relationship between construction procurement strategies and construction contract disputes, Paper presented at the Proceedings of CIB W92 Symposium: North Meets South: Developing Ideas, University of Natal.

David, C. (2011) *Building Contract Claims* (5th ed.), Wiley-Blackwell.

DEVB (2015) *Consultation Document: Proposed Security of Payment Legislation for the Construction Industry*, Development Bureau, Government of the HKSAR.

Diekmann, J., Girard, M., & Abdul-Hadi, N. (1994) Disputes potential index: A study into the predictability of contract disputes. Source Document 101, Construction Industry Institute, University of Texas.

Environment, Transport and Works Bureau (2002) Correction rules for tender errors, Technical Circular (Works) No. 41/2002, Environment, transport and Works Bureau, Government Secretariat.

Heath, B., Hills, B., & Berry, M. (1994) The nature and origin of conflict within the construction process, *CIB Report*, 171:35–48.

Hewitt, R. (1986) *Winning Contract Disputes: Strategic Planning for Major Litigation*, Arthur Young.

HKIA, HKICM, & HKIS. (2005) *Agreement & Schedule of Conditions of Building Contract for Use in the Hong Kong Special Administrative Region – Standard form of Building Contract* (Private ed.), The Hong Kong Institute of Architects, The Hong Kong Institute of Construction Managers, and The Hong Kong Institute of Surveyors.

Hong Kong Housing Authority (2017a) *Guidebook on the Use of HKHA 2013 Conditions of Contract for Building Works*, Hong Kong Housing Authority.

Hong Kong Housing Authority. (2017b) *The Hong Kong housing authority general conditions of contract for building works.*

Hong Kong International Arbitration Centre (2019) Why arbitrate in Hong Kong? Available at: www.hkiac.org/arbitration/why-hong-kong [Accessed 28 December 2020].

Kumaraswamy, M. (1997) Common categories and causes of construction claims, *Construction Law Journal*, 13(1):21–34.

Mitropoulos, P., & Howell, G. (2001) Model for understanding, preventing, and resolving project disputes, *Journal of Construction Engineering and Management*, 127(3):223–231.

Mururu (1991) Anatomy of a dispute, *Arbitration*, 57(4):262–264.

Rhys Jones, S. (1994) How constructive is construction law? *Construction Law Journal*, 10:28–38.

RIBA, Plan of Work (2020) Available at: www.ribaplanofwork.com.

Semple, C., Hartman, F. T., & Jergeas, G. (1994) Construction claims and disputes: Causes and cost/time overruns, *Journal of Construction Engineering and Management*, 120(4):785–795.

Spittler, J. R., & Jentzen, G. H. (1992) Dispute resolution: Managing construction conflict with step negotiations, *AACE International Transactions*, 1, D, 9(1).

The Government of Hong Kong SAR. (1999) *The General Conditions of Building Contract*.

The Government of Hong Kong SAR. (2010) Cap. 609 Arbitration Ordinance, Common law of Hong Kong. Available at: https://www.elegislation.gov.hk/hk/cap609 [Accessed 28 August 2021].

The Government of Hong Kong SAR. (2014) Hong Kong policy address 2014: Support the needy let youth flourish unleash Hong Kong's potential. Available at: https://www.policyaddress.gov.hk/2014/eng/pdf/PA2014.pdf [Accessed 28 August 2021].

The Government of Hong Kong SAR. (2015) Hong Kong policy address 2015: Uphold the rule of law seize the opportunities make the right choices. Pursue democracy boost the economy improve people's livelihood. Available at: https://www.policyaddress.gov.hk/2015/eng/pdf/PA2015.pdf [Accessed 28 August 2021].

The Government of Hong Kong SAR. (2019) *Special conditions of contract clause 59, General conditions of building contract*.

The Hong Kong High Court. (2009) *The Practice direction 6.1, the construction and arbitration list*.

Tillett, G., & French, B. (1991) *Resolving Conflict: A Practical Approach* (Vol. 199), University Press Sydney.

Williamson, O. E. (1979) Transaction-cost economics: The governance of contractual relations. *The Journal of Law and Economics*, 22(2):233–261.

Yates, D. J. (1998) Conflict and dispute in the development process: A transaction cost economics perspective. Available at: www.prres.net/proceedings/proceedings1998/Papers/Yates3Ai.PDF.

Yiu, K., & Cheung, S. (2004) Significant dispute sources of construction mediation, Paper presented at the Proceedings of the 1st International Conference on the World of Construction Project Management, 27–28 May (CD-ROM Proceedings).

Index

Note: Page numbers in *italics* indicate a figure and page numbers in **bold** indicate a table on the corresponding page.